D0794168

Nanotechnology 101

Recent Titles in the Science 101 Series

Evolution 101
Janice Moore and Randy Moore

Biotechnology 101
Brian Robert Shmaefsky

Cosmology 101
Kristine M. Larsen

Genetics 101
Michael Windelspecht

Human Origins 101
Holly M. Dunsworth

NANOTECHNOLOGY 101

JOHN MONGILLO

SCIENCE 101

GREENWOOD PRESS
Westport, Connecticut • London

Library of Congress Cataloging-in-Publication Data

Mongillo, John F.
Nanotechnology 101 / John Mongillo.
p. cm. – (Science 101, ISSN 1931-3950)
Includes bibliographical references and index.
ISBN–13: 978–0–313–33880–9 (alk. paper)
1. Nanotechnology–Popular works. I. Title.
T174.7.M66 2007
620′.5–dc22 2007029805

British Library Cataloguing in Publication Data is available.

Library of Congress Catalog Card Number: 2007029805
ISBN-13: 978–0–313–33880–9
ISSN: 1931-3950

First published in 2007

Greenwood Press, 88 Post Road West, Westport, CT 06881
An imprint of Greenwood Publishing Group, Inc.
www.greenwood.com

Printed in the United States of America

The paper used in this book complies with the Permanent Paper Standard issued by the National Information Standards Organization (Z39.48–1984).

10 9 8 7 6 5 4 3 2 1

CONTENTS

Series Foreword

What should you know about science? Because science is so central to life in the twenty-first century, science educators believe that it is essential that *everyone* understand the basic foundations of the most vital and far-reaching scientific disciplines. *Science 101* helps you reach that goal—this series provides readers of all abilities with an accessible summary of the ideas, people, and impacts of major fields of scientific research. The volumes in the series provide readers—whether students new to the science or just interested members of the lay public—with the essentials of a science using a minimum of jargon and mathematics. In each volume, more complicated ideas build upon simpler ones, and concepts are discussed in short, concise segments that make them more easily understood. In addition, each volume provides an easy-to-use glossary and annotated references and resources of the most useful and accessible print and electronic resources that are currently available.

PREFACE

OVERVIEW

Nanotechnology: A 21st-Century Technology

> Nanotechnology is the ability to observe, measure, manipulate, and manufacture things at the nanometer scale, the size of atoms and molecules.

The National Science Foundation predicts that the global marketplace for goods and services using nanotechnologies will be worth a trillion dollars by 2015. In the same year, career opportunities in this fast-paced technology will require 2–5 million semiskilled and skilled employees worldwide. Nanotechnology has the potential to affect everything from the clothes we wear, to the energy we use, to the way we detect and treat cancer and other diseases.

Nanotechnology 101 is a reference book and an excellent research tool for developing a working knowledge of basic nanotechnology concepts and topics. Besides the potential benefits of nanotechnology, the legal, ethical, and social issues are also addressed.

In recent years, there have been several nanotechnology surveys that have found that a majority of the adults know little or nothing about nanotechnology. Therefore, in an effort to make the book more accessible and appealing to a broad range of readers, *Nanotechnology 101* incorporates many features and enhancements that include:

- **Current Nanotech Topics.** This book provides all of the basic concepts and topics of nanotechnology. In its comprehensive, easy-to-read format, *Nanotechnology 101* covers nanotechnology's influence in medicine, engineering, economics, electronics, communications, energy, environment, transportation, space travel, and homeland security.

- **Bibliography.** An excellent listing of reading materials at the end of each chapter and in the appendix is provided.
- **Interviews**. This book also features more than 15 interviews with people who work and study in the nanotechnology field. The interviews include a nanotechnology business owner, an electrical engineer, a physics professor, an environmentalist, a chemistry teacher, and two high school students who were winners of an NSTA nanotechnology project.
- **Companies and Products**. General information and Web sites for more than 100 companies who are involved in cutting-edge nanotechnology research and nanofabrication have been presented to the reader. Some of these companies include IBM, Intel, Samsung, Eastman Kodak, Dupont, and Hitachi High Technologies.
- **Career Information.** It encourages young people to explore the possibilities of a career in the nanotechnology field by providing resources they can contact to learn more about career opportunities in the field. As an example, Penn State provides a free CD, *Careers in Nanofabrication.*
- **Video Sources.** To enhance the text, *Nanotechnology 101* provides more than 80 nanotechnology video Web sites. As an example, *Is Nanotechnology Going to be the Next Industrial Revolution?* Conversations in Science, Madison Metropolitan School District, UW-Madison Interdisciplinary Education Group. If you would like to view the video go to: http://mrsec.wisc.edu/Edetc/cineplex/MMSD/nano5.html
- **Hands-on Activities and Suggested Projects**. There are suggestions and opportunities for students and teachers to explore nano activities. As an example, students can make a nanometer ruler, construct a Buckyball model, and build a ***LEGO***® model of an atomic force microscope. Want to know how a scanning electron microscope works? Students can access NASA's electron microscope virtual lab Web site.
- **Government and Nongovernment Resources**. All of the major government agencies that are conducting nanotechnology research and funding are included. Familiar names include Department of Energy, Environmental Protection Agency, and NASA. Many nongovernment groups are listed as well. This list includes such organizations as, Foresight Institute, Nanotech Institute, Nano Science and Technology Institute, NanoBusiness Alliance, Institute of Technology, and Nanotech.
- **Nanotechnology Timeline of Events.** The timeline provides an opportunity to trace some of the important events of the history of nanotechnology.
- **University and College Resources.** *Nanotechnology 101* includes nanotech resource links and Web sites from more than 20 colleges and universities that provide nanotechnology educational and outreach programs devoted to middle school or high school students. By contacting these college sites,

students and teachers can access interactive activities, lesson plans, online exhibits, experiments, games, and video broadcasts.

- **National Science Education Standards**. The content in the book provides a close alignment with the *National Science Education Standards*. Nanotechnology is not a traditional discipline, but rather a combination of disciplines involving physics, chemistry, biology, mathematics, engineering, and technology. *Nanotechnology 101* provides information in the appendix that links nanotechnology concepts with those science education standards in each of the major science fields.

RICHARD E. SMALLEY (1943–2005)

To many people, Richard Smalley was the foremost leader in nanotechnology. He has often been noted as the "Father of Nanotechnology."

Richard Smalley was a Rice University professor who won a Nobel Prize in chemistry in 1996. He is well known for his work with carbon nanotubes (known as the "Buckyballs").

In the spring of 1999, Professor Smalley addressed the Senate Subcommittee on Science, Technology, and Space. Testifying at the meetings, he had called for congressional support of a National Nanotechnology Initiative (NNI) that could double U.S. Federal funding for nanoscale research and development efforts over the next 3 years, to about $460 million.

In 1999, while Professor Richard Smalley was testifying before the Congress, he had also been fighting leukemia for over a year. At one meeting, he had the opportunity to speak about the medical applications of nanotechnology. Here is an excerpt of what he had to say.

> I sit before you today with very little hair on my head. It fell out a few weeks ago as a result of the chemotherapy I've been undergoing to treat a type of non-Hodgkin's lymphoma. While I am very optimistic, this chemotherapy is a very blunt tool. It consists of small molecules, which are toxic—they kill cells in my body. Although they are meant to kill only the cancer cells, they kill hair cells too, and cause all sorts of other havoc.
>
> Now, I'm not complaining. Twenty years ago, without even this crude chemotherapy, I would already be dead. But 20 years from now, I am confident we will no longer have to use this blunt tool. By then, nanotechnology will have given us specially engineered drugs, which are nanoscale cancer-seeking missiles, a molecular technology that specifically targets just the mutant cancer cells in the human body and leaves everything else blissfully alone. . . . I may not live to see it. But, with your help, I am confident it will happen. Cancer—at least the type that I have—will be a thing of the past.

Richard Smalley died of non-Hodgkin's lymphoma on October 2005.

In 1999, the Rice University Center for Nanoscale Science and Technology (CNST) was renamed the Richard E. Smalley Institute for Nanoscale Science and Technology in his honor.

Note: Non-Hodgkin's lymphoma (NHL) is the nation's sixth leading cause of cancer death. The American Cancer Society predicted that there would be about 6,300 new cases of NHL in this country in 2007. About 18,000 people die of this disease each year.

Acknowledgments

The author wishes to acknowledge and express the contribution of government and nongovernment organizations and companies. A special thanks to the following people who provided resources, photos, and information: Dr. Laura Blasi at Saint Leo University, Florida; Mr. Paul Octavio Boubion, 8th grade physical science teacher at the Carl H. Lorbeer Middle School in California; high school students, Matt Boyas and Sarah Perrone who attend the Upper St. Clair School District in Pennsylvania; Associate Professor Dean Campbell, Ph.D. at Bradley University in Peoria, Illinois; Dr. Richard Claus, president of NanoSonic Inc. located in Virginia; Dr. Martin L. Culpepper at Massachusetts Institute of Technology in Massachusetts; Renee DeWald, chemistry teacher at the Evanston Township High School in Illinois; Norma L. Gentner, Enrichment Teacher at Heritage Heights Elementary in Amherst, New York; Dr. Nancy Healy, Education Coordinator of the National Nanotechnology Infrastructure Network (NNIN) in Georgia; Catherine Marcotte, science teacher, ForwardVIEW Academy in Rhode Island; Patricia Palazzolo, Gifted Education Coordinator, for the Upper St. Clair School District in Pennsylvania; Dr.Makarand Paranjape at Georgetown University Washington, DC; Edith A., Perez, Medical Doctor and a Professor of Medicine at the Mayo Medical School in Florida; Professor Timothy D. Sands,Ph.D. at Purdue University; Professor Paul G. Tratnyek, Ph.D. at Oregon Health & Science University's OGI School of Science & Engineering; Nathan A. Unterman, physics teacher at the Glenbrook North High School in Illinois; Dr. Alyssa Wise and Dr. Patricia Schank (NanoSense team) and Dr. Brent MacQueen at SRI International in California.

Many thanks to Amy Mongillo and Dan Lanier who provided special assistance in reviewing topics and offering suggestions. In addition, Cynthia Sequin of Purdue Public Relations, Catherine McCarthy, Grants Project Director of Sciencenter, Cornell University, IBM, Konarka Technologies, Oak Ridge National Laboratory, and Hitachi High-Technologies (Japan) who provided technical assistance in acquiring photos for the book.

1

What Is Nanotechnology?

INTRODUCTION

A patron at a local restaurant spills coffee on his trousers. The liquid beads up and rolls off without leaving a spot on his clothing.

The U.S. Golf Association suggests that golfers can now use new golf balls that fly straighter, with less wobble, than normal golf balls.

A woman is using a new variety of canola oil in preparing her meals. The oil contains tiny particles that block cholesterol from entering her bloodstream.

Walking down a street in London, England, a pedestrian suddenly smells that the air is cleaner. The sidewalk is treated with a special product that breaks down harmful pollutants in the air.

The air purifying pavement, the new golf balls, the nonstain pants, are just some of the examples of products produced by nanotechnology, a key technology for the 21st century. Nanotechnology offers cutting-edge applications that will revolutionize the way we detect and treat disease, monitor and protect the environment, produce and store energy, improve crop production and food quality, and build complex structures as small as an electronic circuit or as large as an airplane.

WHAT IS NANOTECHNOLOGY?

Nanotechnology is the ability to observe, measure, manipulate, and manufacture things at the nanometer scale. A nanometer (nm) is an SI (*Système International d'Unités*) unit of length 10^{-9} or a distance of one-billionth of a meter. That's very small. At this scale, you are talking about the size of atoms and molecules.

To create a visual image of a nanometer, observe the nail on your little finger. The width of your nail on this finger is about 10 million

Table 1.1 Some Common Objects in Nanometers

How Many Nanometers?	Approximately
The Width of an Atom	1 nanometer (nm)
The Width Across a DNA Molecule	2 nanometers
The Width of a Wire in a Computer	100 nanometers
The Wavelength of Ultraviolet Light	300 nanometers
The Width of a Dust Particle	800 nanometers
The Length of Some Bacteria	1,000 nanometers
The Width of a Red Blood Cell	10,000 nanometers
The Width of a Hair	75,000 to 100,000 nanometers
The Width Across a Head of a Pin	1,000,000 nanometers
The Width Across the Nail of a Little Finger	10,000,000 nanometers

nanometers across. To get a sense of some other nano-scaled objects, a strand of human hair is approximately 75,000 to 100,000 nanometers in diameter. A head of a pin is about a million nanometers wide and it would take about 10 hydrogen atoms end-to-end to span the length of one nanometer.

The word "nanotechnology" was first introduced in the late 1970s. While many definitions for nanotechnology exist, most groups use the National Nanotechnology Initiative (NNI) definition. The NNI calls something "nanotechnology" only if it involves all of the following:

- Research and technology development at the atomic, molecular, or macromolecular levels, in the length scale of approximately 1 to 100-nanometer range.
- Creating and using structures, devices, and systems that have novel properties and functions because of their small and/or intermediate size.
- Ability to control or manipulate on the atomic scale.

See Chapter 8 for more information about the National Nanotechnology Initiative.

Please note that nanotechnology is not merely the study of small things. Nanotechnology is the research and development of materials, devices, and systems that exhibit physical, chemical, and biological properties. These properties can be different from those found at larger scales—those that are more than 100 nanometers.

Living with Nanoparticles

You live with nanoparticles every day. In a normal room, there are about 15,000 nanoparticles per cubic centimeter in the air. If you are taking a walk in the forest, you will be in an environment where there will be about 50,000 nanoparticles per cubic centimeter. In a large city, there can be about 100,000 nanoparticles per cubic centimeter (cm^3).

NANO, NANO, NANO

In this chapter as well as in the others, we will be using terms, such as nanostructures, nanodevices, nanoparticles, nanoscale, nanomedicine, nanowires, nanotubes, nanoengineering, and so forth. The prefix, "nano," is used to indicate a tool, an enterprise, a particle, a phenomenon, a project, or a manufactured item operating on or concerned with a scale at one billionth of a meter.

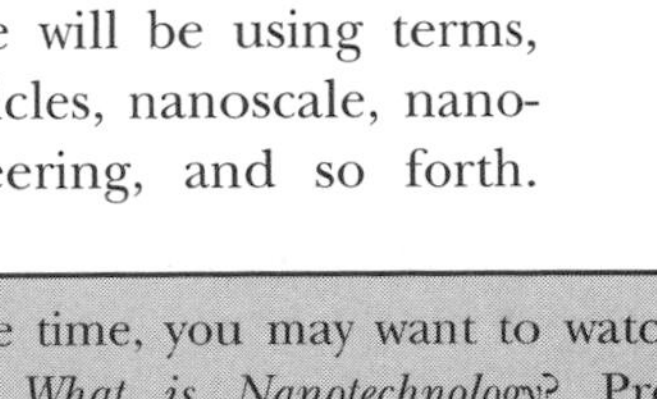

If you have time, you may want to watch the video, *What is Nanotechnology*? Professor Wendy Crone narrates the video. http://www.sciencedaily.com/videos/2006-06-11/

NANOTECHNOLOGY, A FUTURE TRILLION DOLLAR BUSINESS

Making products at the nanometer scale is and will become a big economy for many countries. By 2015, nanotechnology could be a $1 trillion industry. Meanwhile, according to the National Nanotechnology Initiative (NNI), scientists will create new ways of making structural materials that will be used to build products and devices atom-by-atom and molecule-by-molecule. These nanotechnology materials are expected to bring about lighter, stronger, smarter, cheaper, cleaner, and more durable products.

Did you know?
Nano is derived from the Greek word meaning "dwarf."

Did you know?
Even though you cannot see nanometer-sized particles, you can smell some of them. As an example, the particles (molecules) in the air that float from a cake baking in an oven are less than a nanometer in size. They are suspended in the air because gravity does not have much of an effect on them, owing to their small size and mass. As the particles disperse around the room, they reach our noses and we can smell them less than a nanometer away.

NANOTECHNOLOGY WILL DEVELOP IN STAGES

Dr. Mihail Roco is a senior advisor on nanotechnology to the National Science Foundation (NSF) and coordinator of the NNI. Dr. Roco believes that nanotechnology will have several phases of development. He states that we are now in the second phase. The first one consisted of using nanostructures, simple nanoparticles, designed to perform one task. The second phase started in 2005. In the second phase, researchers have discovered ways to precisely construct nanoscale building blocks. The blocks can be assembled into flat or curved structures such as bundles, sheets, and tubes. These structures hold promise for new and powerful drug delivery systems, electronic circuits, catalysts, and light energy sources.

By 2010, Dr. Roco says, the third phase will arrive, featuring nanosystems with thousands of interacting components. A few years after that, the first "molecular" nanodevices will appear. These devices will be composed of systems within systems operating much like a human cell works.

One of the main reasons why there is a lot more activity in producing nanotechnology products today than before is because there are now many new kinds of tools. These new tools consisting of special scanning electron microscopes and atomic force microscopes can measure, see, and manipulate nanometer-sized particles. See Chapter 3 for more information about nanotechnology tools.

NANOTECHNOLOGY PRODUCTS AND APPLICATIONS

What kinds of nanoproducts are available now?

More than $32 billion in products containing nanomaterials were sold globally in 2005. As was stated earlier, the global marketplace for goods and services using nanotechnologies will grow to $1 trillion by 2015. However, other financial experts predict that the marketplace for nanoproducts will reach $2.6 trillion in manufactured goods by 2014.

Presently, there are more than 200 companies that market and sell products using nanotechnology applications. Many of these companies produce much of the 700 or so nanoproducts that are currently available in the U.S. marketplace. Let's review a few examples of nanoproducts that are available today.

Sporting Goods

Special nanoparticles made of carbon are used to stiffen areas of the racquet head and shaft. The particles are 100 times more rigid than steel and 6 times lighter. The new composite hockey stick developed is more

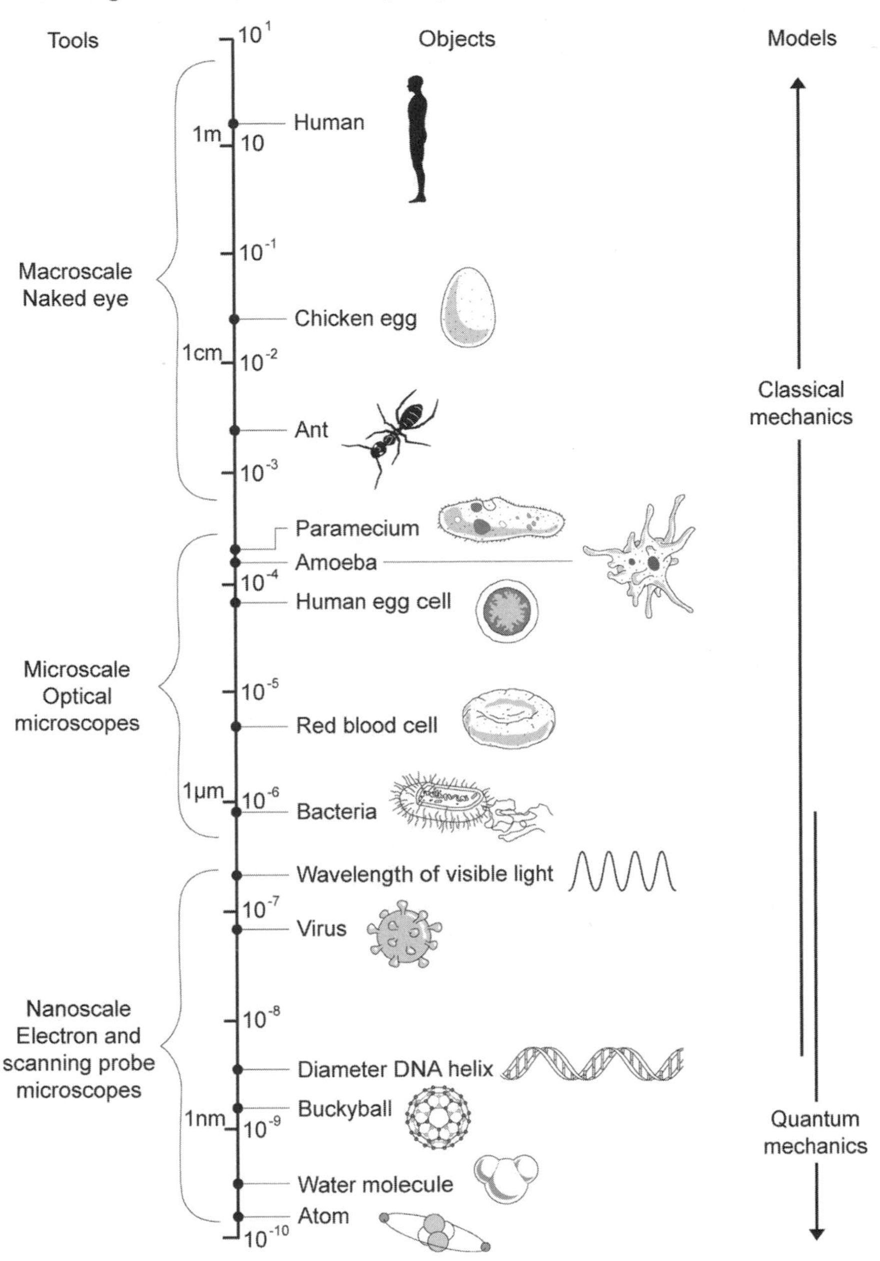

Figure 1.1 Scale of Things. (*Courtesy of Jeff Dixon*)

durable than other sticks because of the carbon nanotube epoxy matrix. Nano tennis balls coated internally with a nano-sized membrane, slow pressure drain without adding weight.

Car Paint and Car Waxes

There are now new kinds of automobile paints, developed from nanotechnology, that have improved scratch-resistant qualities compared to conventional car paint. Nano car waxes, made with nano-sized polishing agents, provide a better shine due to its ability to fill-in tiny blemishes in automotive paint finishes.

Antibacterial Cleansers

There are several antibacterial cleansers that use nanoemulsion technology to kill pathogens. The cleansers are able to kill tuberculosis and bacteria while remaining nonflammable, noncorrosive, and nontoxic. The good news is that there are no harmful effects when using these products.

Medical Bandages

Silver's antibiotic properties have made the metal a popular treatment for wounds and burns. Special dressings for burns provide antimicrobial barrier protection using concentrations of nano silver particles. These medical bandages help skin to heal by preventing infections during treatment. The silver-impregnated dressings require fewer painful changings of dressings than previous silver treatments.

> **Did you know?**
> The cleansing power of silver was known in the days of the Roman Empire, when silver coins were used to purify water in jugs and containers for drinking and cooking purposes.

Apparel Industry

Several clothing companies have marketed new brands of nonstain nanotechnology fabrics. The fabric resists spills from many types of fibers (cotton, synthetics, wool, silk, rayon, and polypropylene). The fabric also repels a range of liquids including beverages and salad dressings. These fabrics keep the body cool and comfortable and have an antistatic treatment that reduces static cling from dog hair, lint, and dust. See Chapter 6 for more information about nanotechnology fabrics.

Sunscreens and Cosmetics

Several cosmetic companies have already marketed a number of products that include nanotechnology. The products include sunscreens, deodorants, and antiaging creams. One sunscreen product uses zinc oxide nanoparticles that protects the skin but does not leave white marks on the skin. However, more regulation of some of these products will be needed because of the limited research that has been carried out in this still new field. See Chapter 6 for more information about sunscreens.

Organic Light-Emitting Displays or OLEDs

The OLEDs are ultrathin displays manufactured by sandwiching extremely thin (often nano-sized) layers of organic polymer light-emitting materials between electrodes. These images are bright and viewable at wide angles. The displays are smaller and of lighter weight than traditional LCD (liquid crystal display) displays—meaning they are ideally suited to mobile electronics, such as digital cameras, cellular phones, and handheld computers.

FUTURE APPLICATIONS OF NANOTECHNOLOGY

What can we look for in future applications of nanotechnology?

Environment

Emerging nanotechnologies hold great promise for creating new means of detecting air pollutants, and cleaning polluted waste streams and groundwater. A research group is now testing the use of magnetic nanoparticles that can absorb and trap organic contaminants in water. If the testing continually succeeds, the process can also be very effective in cleaning up contaminated Superfund sites—hazardous and toxic waste sites—in the United States. See Chapter 8 for more information about the environment.

Solar Energy

Researchers are making an effort to find a lower-cost source of household energy for the nation's future. They are exploring the creation of nanoscale devices on the molecular level that can convert sunlight into electric current. Scientists have invented a plastic solar cell that can turn the sun's power into electrical energy, even on a cloudy day. The plastic material uses nanotechnology and contains solar cells able to harness the sun's invisible, infrared rays. Like paint, the composite can be sprayed onto other materials and used as a portable source of electricity. A sweater coated in the material could power a cell phone or other wireless devices.

However, the idea of using nanostructures to convert sunlight into electricity is still theoretical. The main concern is that the present cost of electricity produced by solar cells is four times greater than electricity produced by nuclear or fossil fuels. Many companies and government agencies are funding much research in solar projects, which indicates there is an increasing interest in this field by the scientific community and corporations. See Chapter 8 for more information about solar cells.

Fuel Cells

Several companies are utilizing nanostructure technology to help develop high performance fuel cells for use in automobiles, portable consumer electronics such as laptop computers, cell phones, and digital cameras. A fuel cell is an energy conversion device and alternative to batteries that converts energy from a chemical reaction into electricity and heat. Fuel cells combine fuels such as hydrogen or methanol along with air and water to produce electrical power. Because their by-products are heat and water, fuel cells are environmentally friendly. See Chapter 8 for more information about fuel cells.

Food and Agriculture

The ability to use nanotechnology will allow food companies to design and provide food products that would be safer, cheaper, and more sustainable than the foods today. Food companies will also use less water and chemicals in the preparation and production of food products.

One food company had developed nanosensors that would be embedded in food packaging. A color change in the nanosensor would alert the consumer if a food in the package had become contaminated or if it had begun to spoil. Some companies are producing a plastic consisting of clay nanoparticles. The nanoparticles in the plastic are able to block out oxygen, carbon dioxide, and moisture from spoiling fresh meats and other foods. See Chapter 7 for more information about food and agriculture.

Automobiles and Aeronautics

Nanoscale powders and nanoparticles will be able to enhance the physical properties of automobile, aircraft, watercraft, trains, and spacecraft. Planes, trains, and automobiles will be lighter, faster, and more fuel-efficient and constructed of lighter, stronger materials. Some of these lightweight materials will include aluminum bodies for automobiles, brake systems for high-speed trains, and quieter aircraft engines. The stronger, lighter materials will help in energy efficiency and reducing mass and weight of finished products. See Chapter 6 for more information about automobiles and aircraft.

Medical Applications

Many medical procedures could be handled by nanomachines that rebuild arteries, rebuild bones, and reinforce bones. In cancer nanotechnology research, scientists are testing and experimenting with new ideas to diagnose, treat, and prevent cancer in the future.

One research medical team is using nanoshells to target cancer cells. Nanoshells are hollow silica spheres covered with gold. In animal testing, Naomi Halas's research team at Rice University directed infrared radiation through tissue and onto the shells, causing the gold to superheat and destroy tumor cells while leaving healthy ones intact. Human clinical trials using gold nanoshells are slated to begin within a few years.

Another cancer research team has shown that the targeted gold nanoparticles combined with lasers can kill oral cancer cells. Oral cancer is any cancerous tissue growth located in the mouth. Smoking and other tobacco use are associated with 70–80 percent of oral cancer cases. Thirty thousand Americans will be diagnosed with oral or pharyngeal cancer this year. See Chapter 5 for more information about the medical field.

Lab-on-a-Chip

In other nano medical news, researchers are studying lab-on-a-chip technology. Lab-on-a-chip technology consists of a portable handheld device containing a simple computer chip that can diagnose and monitor the medical conditions of a patient. As an example, a tiny sample of blood placed on the device could diagnose if the patient is diabetic. The lab-on-a-chip could be used for commercial, medical diagnostic applications, such as an in-office test for strep throat, or modern in-home pregnancy tests.

NASA has customized lab-on-a-chip technology to protect future space explorers. The lab-on-a-chip would be used to monitor the health of the crew by detecting contaminants in the spacecraft.

THE U.S. GOVERNMENT INVESTS IN NANOTECHNOLOGY RESEARCH

Nanotechnology is expected to have large impact on many sectors of the world's economy. A strong nanotechnology economy can lead to new products, new businesses, new jobs, and even new industries for many countries. As a result, nanotechnology funding for research is growing rapidly all over.

In 2001, President Clinton requested a major new initiative in the 2002 federal budget, called the National Nanotechnology Initiative (NNI). The budget included an increase of more than 200 million dollars for the government's investment in nanotechnology research and development. In December 2003, President Bush signed the Nanotechnology Research and Development Act, which authorized funding for nanotechnology research and development (R&D) over 4 years, starting in FY 2005. This legislation puts into law programs and activities supported by the National Nanotechnology Initiative.

The 2003 bill authorized nearly $3.7 billion for research and development programs coordinated among several federal agencies. The legislation also authorized public hearings and expert advisory panels, as well as the American Nanotechnology Preparedness Center to study the emerging technology's potential societal and ethical effects.

Other Countries Are Also Investing in Nanotechnology Research

The United States is not the only country to recognize the tremendous economic potential of nanotechnology. In 2007, the Russian government planned to take its export revenues from oil and gas sales and invest it in nanotechnology research and development. During the next 3 years, Russia will invest more than $1 billion in nanotechnology so that it can become less dependent on raw materials.

Brazil had a projected budget for nanoscience during 2004 to 2007 that included creating three institutions, four networks, and recruiting 300 scientists to work in nanotechnology research. Some of the Brazilian researchers are interested in the use of magnetic nanoparticles to remove oil from oil spills and then be able to recycle the nanoparticles and the oil. Other countries pursuing nanotechnology include Thailand, Philippines, Chile, Israel, Mexico, Argentina, South Africa, Japan, China, and Korea, as well as several European countries.

WHAT DO AMERICANS THINK OF NANOTECHNOLOGY?

So, what do Americans think about this technology? In 2004, North Carolina State University researchers conducted a survey designed to find out about the public's perceptions of nanotechnology. The telephone survey polled a random sample of 1,536 adults in the continental United States in the spring of 2004 as a part of a larger research project studying public perceptions of nanotechnology. What follows are some of their findings.

- More than 80 percent of the adults indicated they had heard "little" or "nothing" about nanotechnology. Most of them could not correctly answer factual questions about it. However, despite knowing very little about the science, 40 percent of the respondents predicted nanotechnology would produce more benefits than risks. Another 38 percent believed risks and benefits of nanotechnology would be about equal, and only 22 percent said the risks would outweigh the benefits.
- Approximately 70 percent of those surveyed said they were "somewhat" or "very" hopeful about nanotechnology, while 80 percent said they were not worried at all about the science. Only 5 percent said they felt angry about the science.
- Respondents were also asked to choose the most important potential benefit from nanotechnology. A majority (57 percent) cited "new and better ways to detect and treat human diseases." Sixteen percent selected "new and better ways to clean up the environment"; 12 percent chose "increased national security and defense capabilities"; and 11 percent identified ways to "improve human physical and mental abilities" as the most important benefit.
- In choosing which potential risk was the most important to avoid, most respondents (32 percent) picked "losing personal privacy to tiny new surveillance devices." Others wanted to avoid "a nanotechnology inspired arms race" (24 percent); "breathing nano-sized particles that accumulate in your body" (19 percent); "economic disruption caused by the loss of traditional jobs."

For the full report, *Study Shows Americans Encouraged by Prospects of Nanotechnology*, contact North Carolina State University.

WILL NANOTECHNOLOGY BE USED TO HELP PEOPLE IN DEVELOPING COUNTRIES?

The applications of nanotechnology will certainly benefit all the developed countries. But will this new technology help the developing countries—those nations that have limited resources and whose people live in poor conditions? To answer the question, the Canadian Program on Genomics and Global Health (CPGGH) has a plan to assist those developing countries.

The GPGGH is a leading international group that studies nanotechnology applications. They advocate that nanotechnology applications will help people in developing countries tackle their most urgent problems. Some of these problems include extreme poverty and hunger, child mortality, environmental degradation, and diseases, such as malaria and HIV/AIDS.

The goals of the CPGGH identified and ranked several nanotechnology applications most likely to have an impact in the developing world. The top five nanotechnology applications on the list included:

1. Energy storage, production, and conversion
2. Agricultural productivity enhancement
3. Water treatment and remediation—cleaning up waste sites
4. Disease diagnosis and screening
5. Drug delivery systems

The CPGGH group selected energy as the number one priority in nanotechnology applications to assist developing countries. They included energy production, conversion and storage, along with creation of alternative fuels, as the energy areas where nanotechnology applications are most likely to benefit developing countries.

Number two on the list is agriculture. They state that science can be used to develop a range of inexpensive nanotech applications to increase soil fertility and crop production, and help eliminate malnutrition. Malnutrition is a contributor to more than half the deaths of children under five in developing countries. Other agricultural developments include nanosensors to monitor the health of crops and farm animals and magnetic nanoparticles to remove soil contaminants.

Water treatment is third-ranked by the panel. One-sixth of the world's population lacks access to safe water supplies, according to their study group. More than one-third of the population of rural areas in Africa, Asia, and Latin America has no clean water. Two million children die each year from water-related diseases, such as diarrhea, cholera, typhoid, and schistosomiasis, which result from a lack of adequate water sources and sanitation. Nanomembranes and nanoclays are inexpensive, portable, and easily cleaned systems that purify, detoxify, and desalinate water more efficiently than conventional bacterial and viral filters. These water treatment processes could provide potable water for families and communities.

Disease diagnosis and screening was ranked fourth. Some of these technologies include inexpensive, handheld diagnostic kits that detect the presence of several pathogens at once and could be used for wide-range screening in small clinics.

Drug delivery system was fifth on the list. Nanotechnology could improve transportation costs by developing drugs to last longer in storage. Most drugs today have a short shelf life because of changes in humidity and temperature change in storage rooms.

THE NANOTECHNOLOGY JOB MARKET

The Need for Workers

Many jobs will be needed to fill in the vacancies for nanotechnology. The National Science Foundation ("NSF") projects that the nanotechnology job market in the United States will require over 2 million nanotechnology-savvy workers by 2014. The NSF therefore is calling for children between the ages of 10 and 17 to be educated now about the field that will define their job market as adults. Of the 2 million nanotechnology-savvy workers required by 2014, 20 percent are expected to be scientists, with the remaining 80 percent consisting of highly skilled engineers, technicians, business leaders, and economists.

Dr. Mihail Roco, mentioned earlier in this chapter, states that as nanotechnology moves into the mainstream, companies building products at the atomic level eventually will face a serious shortage of talent—far worse than what is already occurring. Dr. Roco estimated, about 2 million nanotech-trained workers will be needed to support growing industries and the startups they spawn within the next 10 to 15 years, He emphasized that the country needs to find ways to motivate students about sciences and to make them aware of the career opportunities in nanotechnology fields.

UNIVERSITIES OFFER NANOTECHNOLOGY YOUTH PROGRAMS

To motivate students to explore careers in nanotechnology, many colleges and universities have presented a variety of programs for middle school and high school students. These schools offer nanotechnology camps, school outreach programs, field trips, and nanotechnology career days for students from city, suburban, and private schools. Some of these universities and colleges include Georgia Institute of Technology, Penn State, University of California, Santa Barbara, Cornell, University

Students interact with the LEGO® nanotechnology exhibit that is one of several displays in the "Nanotechnology: The Science of Making Things Smaller" project. The project is directed by Purdue University's Department of Physics and its School of Electrical and Computer Engineering in conjunction with the Children's Museum of Oak Ridge in Oak Ridge, Tennessee. (*Courtesy Purdue News Service photo/David Umberger*)

of New Mexico, Stanford, Howard University, Michigan State, University of Pennsylvania, Drexel University, Harvard, and the University of Albany.

Let's look at one example of a university's nanotechnology program for young people. The mission at the Northwestern University-Nanoscale Science and Engineering Center is to teach people of all ages about the nanoworld. In one of their workshops, 37 fifth graders came to Northwestern to participate in "NanoDay," a half-day of activities designed to spark student interest in nanoscience and technology.

Besides the universities, the National Nanotechnology Infrastructure Network (NNIN) also provides a variety of school programs. Even museums such as the Exploratorium in San Francisco, the Lawrence Hall of Science in California, and the Museum of Science in Boston have had nanotechnology exhibits. Purdue University and Cornell University

The National Nanotechnology Infrastructure Network provides a variety of school programs. In one of their programs, students from an elementary school are learning about self-assembly through an activity demonstrated by Richard Kirk CEO of Claro Chemicals. (*Courtesy National Nanotechnology Infrastructure Network*)

have sponsored on-the-road nanoexhibits to spark public interest in nanotechnology.

Several high schools also have special nanotechnology programs. The Future is NEAR (Nanotechnology Education and Research) program at North Penn High School in Lansdale, Pennsylvania offers its students an opportunity to gain 21st-century skills that will help prepare them to become successful leaders in the new, technological global society.

In Albany, New York, eligible students at the Albany High School will have the opportunity to participate in nanotechnology education via a pilot program with the University of Albany, College of Nanoscale Science and Engineering. Under its pilot phase, NanoHigh will focus on school-to-work activities designed to train AHS students in creative nanoscience and nanoengineering concepts, and help equip them with the skill set necessary to pursue advanced educational opportunities in the field that is "leading to the next industrial revolution." See Chapter 10 to learn more about how schools, universities, colleges, and other organizations are encouraging young people to explore nanotechnology.

THE FIELDS OF STUDY THAT INFLUENCE NANOTECHNOLOGY

Anyone who explores a career in nanotechnology will learn that it is a multidisciplinary field. Some of these fields include physics, mechanical engineering, chemistry, biology, medicine, business and economics, agricultural science, electronics, computer science, environmental science, and law and ethics.

Chemistry, biology, and physics are the major science nanotechnology fields of study. These fields can provide candidates with a solid foundation for any of a broad range of nanotechnology careers.

Chemistry is the study of matter and the changes matter undergoes. Nanotechnology uses chemistry extensively in the field of molecular devices and molecular manipulation.

The chemistry in nanotechnology is a bit different from the traditional chemistry courses. The chemistry in nanotechnology deals with the manipulation of atoms and molecules of chemicals. Synthesizing a chemical with nanotechnology means building something atom by atom. This process is quite different from traditional chemistry that usually studies materials at the macro or larger scale—more than 100 nanometers.

The traditional *biology* courses involve the study of the molecular structure and function of living systems. These systems range from cells and bacteria to spiders and humans. All of these organisms rely on nanometer-sized proteins or molecular motors to do everything from cell division to moving a leg. An application of nanotechnology in biology would be to isolate one of these molecular motors from a living system and then design and use it to construct a nanoscale product such as a nanorobot to search out and destroy tumors in the human body.

Physics is the study of the most fundamental interactions between energy and matter and time and space. Physics also includes quantum mechanics, the study of matter and radiation at an atomic scale.

Much of modern technology, such as the transistor, and integrated circuits, which are at the heart of modern computers and electronics, is the result of physics research.

To study science and technology, one must have an adequate background in *mathematics.* Basic calculus is essential, and differential equations and linear algebra are widely used.

The field of *engineering* constitutes applied science or the use of scientific principles to analyze and solve practical problems, and especially to design, build, operate, and maintain many types of constructions. The principle fields of engineering are civil and environmental, chemical, electrical, and mechanical engineering. But many other specialized types of engineering exist. All types of engineering involve the application of physics and mathematics to designing or problem solving.

Feature: Mechanical Engineering

Physics is the foundation for all types of engineering. One of the fields is mechanical engineering, a discipline that involves the application of principles of physics for analysis, design, manufacturing, and maintenance of mechanical systems. Mechanical engineering requires a solid understanding of key concepts including mechanics, kinematics, thermodynamics, and energy.

Practitioners of mechanical engineering, known as mechanical engineers, use these principles and others in the design and analysis of automobiles, aircraft, heating and cooling systems, buildings and bridges, industrial equipment and machinery, and more. At the scale of molecular technology, mechanical engineering develops molecular machines and molecular machine components. See the Nano Interview at the end of the chapter for more information about mechanical engineering and its role in nanotechnology.

MAJOR NANOTECHNOLOGY CAREER AREAS

According to the National Nanotechnology Infrastructure Network, opportunities for careers in these fields are expanding rapidly. A major challenge for the field is the education and training of a new generation of skilled workers.

Nanotechnology job projections are estimated to be nearly 2 million workers worldwide by 2015. Most of the jobs will be in the United States, Japan, and several European countries. In addition, nanotechnology will create another 5 to 7 million jobs worldwide in support fields and other industries.

The education levels of this workforce will include skilled workers who have backgrounds in technical programs, have Associates degree (2 years), Bachelors degree (4 years), Masters degree (6 years), and Doctorate (8 years). Note: Not everyone working in the field of nanotechnology will require a doctorate degree. The major career opportunities for these workers will exist in areas such as:

- Electronics/semiconductor industry
- Materials science including textiles, polymers, and packaging
- Automotive and aerospace industries
- Sports equipment
- Pharmaceuticals
- Biotechnology
- Medical fields
- Environmental monitoring and control
- Food science including quality control and packaging
- Forensics—applied sciences used for legal investigations
- University and federal lab research
- National security

However, the nanotechnology fields that will grow most rapidly in the next decade will include:

- Medicine: diagnostics and therapeutics (e.g., drug delivery)
- Energy: fuel cells and batteries
- Robotics, many kinds
- Manufacturing: self-assembly
- Commerce: Radio Frequency Identification
- Space exploration

RICHARD FEYNMAN (1918–1988)

Richard P. Feynman was an American physicist and one of the three men who received the Nobel Prize in Physics in 1965. Feynman, who was a long-term professor at California Institute of Technology, is best known as one of the people who worked on atomic weapons at Los Alamos during World War II. He also discovered what caused the 1986 Challenger explosion.

The idea of atom-by-atom construction was first put forth, in a scientific manner, over 40 years ago by the Nobel Prize-winning physicist Richard Feynman. In 1959, Feynman gave a lecture called "There's Plenty of Room at the Bottom." In his speech, he predicted that an entire encyclopedia would one day fit on the head of a pin and a library with all the world's books would fit in three square yards! Feynman also predicted that microscopic computers, atomic rearrangement, and nanomachines could float in veins.

RICHARD E. SMALLEY (1943–2005)

To many people, Richard Smalley was the foremost leader in nanotechnology. He has often been noted as the "Father of Nanotechnology." Richard Smalley was a Rice University professor who won the Nobel Prize in chemistry in 1996. Richard Smalley is mostly known for his work with carbon nanotubes, (known as the "Buckyballs"). Smalley was hopeful that nanotechnology could solve the global energy problem, which would ultimately solve other worldwide problems such as hunger and water shortages. He believed the potential for nanotechnology to benefit humanity was virtually limitless, and he abided by the mantra: "Be a scientist; save the world."

> **Did you know?**
> Nanoengineered carbon is the most common material used in the nano products followed by silver, a metal, and silica, a compound composed of silicon and oxygen.

In 1999, the Rice University Center for Nanoscale Science and Technology (CNST) was renamed the Richard E. Smalley Institute for Nanoscale Science and Technology in his honor.

WHAT ARE THE RISKS OF NANOTECHNOLOGY?

The long-term benefits of nanotechnology will lead to new means of reducing the production of wastes, cycling industrial contamination, providing potable water, and improving the efficiency of energy production and use. Good news for all. However, there are concerns that the introduction of large quantities of nanostructured materials, such as nanoparticles, into our everyday life may have ethical, legal, and societal consequences.

Two independent studies released in 2003 included the risks of using carbon nanotubes. Chiu-Wing Lam of NASA's Johnson Space Center led one of these studies and the other was led by David Warheit of Dupont. Each study found that carbon nanotubes, when directly injected into the lungs of mice, could damage lung tissue.

One overseas company that makes cleaning products using nanometer-scaled materials had to remove these products from stores because several consumers were getting respiratory problems from using the product.

NANOSAFE2

Several organizations are developing programs to make sure there is a check and balance procedure to insure safety in the nanotechnology field. One European organization consisting of members from seven European Union countries has formed an organization called NanoSafe2. The overall aim of NANOSAFE2 is to develop risk assessment programs and to establish ways to detect, track, and characterize nanoparticles. These methods will be useful in determining any possible risks to humans and the environment. NANOSAFE2 will develop technologies to cut down on exposure to nanoparticles and leaks to environment by designing safe production equipment.

There are other organizations and associations engaged in nanotech safety research. Some of these agencies include the National Science Foundation (NSF), the National Toxicology Program of the Department of Health and Human Services, the Department of Defense (DOD), the Environmental Protection Agency (EPA), the Department of Energy (DOE), and the National Institutes of Health, the National Institutes of Standards and Technology (NIST), and the National Institute of Occupational Safety and Health (NIOSH).

National Institute for Occupational Safety and Health (NIOSH)

NIOSH is conducting research on nanotechnology and occupational health. Its mission is to help answer questions that are critical for supporting the responsible development of nanotechnology.

Nanotechnology Environmental and Health Implications (NEHI)

The Nanotechnology Environmental and Health Implications (NEHI) working group, composed of several U.S. government agencies, has researched and reported on concerns about the safety of nanotechnology products. As of late 2006, one of the overriding concerns NEHI identified was the difficulty and critical importance of reviewing the safety of nanotechnology quickly enough to keep pace with the rapid development and implementation of nanoproducts. Research and development, known as "R and D," can occur very rapidly in the field of nanotechnology, but the study of potential risks of new products to health and the environment can take considerable time to reach conclusions and make decisions about product safety.

National Nanotechnology Initiative (NNI)

The National Nanotechnology Initiative coordinates where federal nanotech research dollars go. About $106 million of the funding was earmarked for environment and health and safety research in nanotechnology.

Environmental Protection Agency (EPA)

In late 2004, the U.S. Environmental Protection Agency awarded about $4 million to fund a dozen research studies at universities on possible risks associated with nanotechnology. See Chapter 8 for more information about the Environmental Protection Agency.

Center for Biological and Environmental Nanotechnology (CBEN)

One of the aims of Rice University's Center for Biological and Environmental Nanotechnology (CBEN) is to characterize the unintended consequences of nanotechnology, particularly in the environmental arena. As one of the six nanoscience and engineering centers funded by the National Science Foundation, CBEN has a mandate to clear major roadblocks to nanotechnology commercialization. As the only center funded under the National Nanotechnology Initiative that focuses exclusively on environmental and biological systems, they are developing ways to use nanotechnology to clean our environment and improve public health.

The Center for Responsible Nanotechnology

The Center for Responsible Nanotechnology has identified several separate and severe risks. Some of their concerns are:

- Economic disruption from an abundance of cheap products.
- Economic oppression from artificially inflated prices.
- Personal risk from criminal or terrorist use.
- Personal or social risk from abusive restrictions.
- Social disruption from new products/lifestyles.
- Unstable arms race.
- Collective environmental damage from unregulated products.
- Free-range self-replicators (grey goo).
- Black market in nanotech.

NANO INTERVIEW: PROFESSOR MARTIN L. CULPEPPER, PH.D., MASSACHUSETTS INSTITUTE OF TECHNOLOGY

Dr. Martin L. Culpepper is a mechanical engineer and Associate Professor at Massachusetts Institute of Technology in Cambridge, Massachusetts. Dr. Culpepper talked with the author about his career in mechanical engineering and his work in nanotechnology.

What is mechanical engineering?

The classical definition of mechanical engineering is the discipline, the science and technology that are required to create mechanical devices, most often those that move themselves or other things. For instance, mechanical engineers design and build mechanical systems that include parts of airplanes, parts of automobiles, and even parts of satellites and spacecraft.

My work, as a mechanical engineering researcher, is to generate the knowledge that is required to build machines and their parts on a small scale, that is, of a size that is approaching the molecular scale. I want to be able to engineer these small-scale machines so that they may be used to move other smaller parts around. For instance, we want to be able to engineer nanomechanisms, nanoactuators and nanosensors; and these machines will all depend upon some sort of nanoscale mechanical device.

Most of mechanical engineering for the last 100 years or so, has dealt with creating machines that were large enough so that we could handle them with our hands. In the last 20 to 30 years engineers have started building devices that are small enough so that one cannot use their hands to handle or operate them, but yet the machines may still be observed via a magnifying glass. When we discuss mechanical engineering in a nanotechnology context we are talking about building machines out of individual atoms, individual molecules, or groups of molecules. The machines I am working on are so small that one cannot see them with a high-powered optical microscope.

There is a great deal of research that must be done before one can engineer nanoscale machines. For instance, we need to understand how to put them together. Once you've made them, how do you take measurements so that you know if they are the right size? How do you measure their motions so that you know if they are moving in the right way? Then, after that, how do you take what you have made and turn it into products that people would use in their everyday lives? As engineers, we do these things with large-scale machines every day and we don't think twice about how to do them. We know how to do this from past experience. At the nanoscale, we struggle to conceive of methods we could use to do the same types of things.

What college did you go to? What was your major?

For my undergraduate studies, I received a scholarship to go to Iowa State University. I was born in Iowa and lived there my entire childhood. While attending Iowa State I worked at John Deere in the summer and I worked in a research group on a project for NASA. From these experiences I became interested in doing research on how to build new machines that people had not been able to engineer before. At the time I graduated from Iowa State, I was not interested in nanotechnology, as it was not as well known then as it is today.

After I graduated, I applied to, and was accepted at, the Massachusetts Institute of Technology (MIT) where I received a Master's of Science degree and then a Ph.D. in mechanical engineering. I am now an Associate Professor and faculty member in Mechanical Engineering at MIT.

What interested you in seeking a career in mechanical engineering?
I became interested in mechanical systems and physics early on. While attending high school, I had pulled apart an automobile carburetor from my dad's car. In those days, the automobile used a carburetor to mix gasoline and air for the engine. Although I was not successful in putting the pieces of the carburetor back together, I was fascinated by the complexity of the parts in the carburetor and wondered how people could design and make things that were so intricate.

Now as an Associate Professor of mechanical engineering at the Massachusetts Institute of Technology, I do research on building intricate, small-scale machines out of atoms and molecules. This work requires an understanding of mechanical engineering and the physics that govern behavior at the molecular scale.

How did you get started in nanotechnology?
About 5 years ago, I was building machines at MIT that were macro-scale in size. These machines were the size of a baseball or football. The macro machines could move precisely and accurately enough so that they could be a type of "construction equipment" for nanoscale machines. They could interact with atoms and measure or help to make nanomachines. Through making the larger machines, I gained knowledge of how one could engineer smaller machines out of atoms and molecules. Now I am designing, building, and testing nanoscale machines.

What is a nanomachine?
Nanoscale devices that can move and push upon other machines or parts. Most biological systems, for instance you and I, are comprised of billions of nanomachines. For instance, our muscle cells act in ways that are similar to macro-scale machine behavior. Within our cells, molecules perform a very intricate manipulation of atoms/other molecules that help our cells replicate and/or heal. Even bacteria have nanoscale motors within them that they use to move around.

The nanomachines that I am interested in building are at an early stage of development. When you look at them, they would resemble the simplest type of machines that you would find at the large scale. For instance, the simpler types of pieces that are put together to build a more complex machine such as bolts, nuts, springs, and structural components. Something with a complex geometry, such as an automobile engine or a robot, will take decades to build at the nanoscale.

There is ongoing work that aims to use carbon nanotubes (tubes that are several atoms in diameter to hundreds of atoms in diameter) to make small-scale machines. Researchers have discovered that these tubes may be used to make bearings similar to those that enable a bicycle wheel rotate. Bearings are one of the fundamental things that you need in order to build machines that move and so engineers are trying to use carbon nanotubes in this way. Scientists and engineers are just getting to the point where they are starting to integrate these small-scale parts in order to build simple nanoscale machines, such as a very small shifting device on a bicycle. Most of these shifting devices rely primarily upon the function of a mechanism, called a 4-bar linkage that shifts the chain between sprockets. Building a shifting device at the nanoscale requires 5 to 10 years of research and trial and error. More complex nanoscale machines, such as robotic grippers, will probably take 10 to 20 years or more.

My group grows the carbon nanotubes and we use simulations to help us understand how to use these tubes to engineer simple machines, for instance the shifter as previously described.

What are some of the benefits of building nanomachines?

Speed and cost are two major benefits of nanomachines. With respect to speed, there are physical laws that govern the speed at which machines may move as a consequence of their mass and stiffness, both of which are usually related to their size. In most cases, the smaller the machine, the faster the machine can be run. If you can build mechanical devices that can run at millions of motions per second, they might be used to accomplish many tasks in a short time. With respect to cost, it becomes inexpensive to build a nanomachine if one uses the right nanomanufacturing processes, because thousands or millions of machines may be made simultaneously.

It is likely that future nanomachines will have an impact on many devices including computers, sensors, and electromechanical devices. It is hoped that nanomachines will increase the speed of store-bought conventional products by a factor of 100 to 1,000. The costs for these new products would also be less than the cost of conventional products.

In addition, smaller machine elements that are made of carbon nanotubes could be used as energy sources for products. The special properties of the nanotubes will allow engineers to build better batteries that have huge energy densities. These batteries may be able to power automobiles, and possibly provide longer-lived portable power sources, for example in an iPod.

Are there any environmental issues or risks in building nanomachines?
One of the things that people should keep in mind is that nanoparticles are small compared to a human cell, roughly 10 – 1000 times smaller. It has been proposed, and I stress the word proposed, that certain types of nanoparticles might be get into cells and do harmful things. I am not an expert in the biology, chemistry or toxicology but I know of experts in these fields that have researched particles to see if they are harmful. In my laboratory, we treat nanoparticles as though they are harmful even though a lot of the evidence shows that they are not. For instance, most people do not know that many cosmetics use certain types of nanoparticles and they have been used for many years without adverse effects.

You mentioned in an article, that potential young scientists needed to learn the language of mathematics. You stated that mathematics is the language of logical thought. Could you expand on that comment?
What I was trying to say is that mathematical expressions are a way to express a logical idea. Many students forget this and then in their eyes, math becomes something abstract that has little use in everyday life. They don't realize that when one thinks about problems, as an engineer or everyday life, they use logic to solve those problems. An engineer or scientist needs to be able to convert common sense and logical ideas in a concise and universally understood form. Mathematics allows us to do this and this helps us to solve engineering and everyday problems. Knowing math is a powerful tool.

What other advice would you give young people who would like a career in mechanical engineering?
My advice is to find somebody who has the same interests that you have, and who seeks excellence in all that they do. Choose someone who can pass those interests on to you and teach you the skills/discipline/fun of doing things well. My parents and grandparents filled these roles for me. They were passionate about the idea of "if you were going to do

something, then do it right." My grandfathers had a passion for working on mechanical things and those skills they had were like magic to me. I wanted to learn to do what they could do. They also did an excellent job and by observing their habit of excellence, I came to expect a lot from myself whenever I do something. Young people should also find a mentor that can pass on to them an understanding of why things are important and why they are relevant to everyday life. At the same time, make sure you have fun doing what interests you. For instance, if you love music or acting, you can learn mechanical engineering and/or physics and then apply them to make better instruments or devices for sets. The same thing applies to people that love cars, robots, airplanes, medicine, woodworking, and sports. You can use mechanical engineering, math, and physics to make useful contributions in these areas and have great fun at the same time.

Would you want any of our readers (teachers/students) to contact you at your lab or by e-mail?
My lab's Web site is http://pcsl.mit.edu

E-mail (culpepper@mit.edu) is OK, but I can't guarantee a response due to the fact that I receive about 100–200 non-spam e-mails per day that I have to respond to.

NANO ACTIVITY: CUTTING IT DOWN TO NANO

The activity, "Cutting It Down to Nano," is adapted from the education team at MRSEC (Materials Research Science and Engineering Center) at the University of Wisconsin.

The challenge of the activity is to determine the number of times you need to cut the strip of paper in half in order to make it between 0 and 10 nanometers long. See the following Web site for more information about the activity: mrsec.wisc.edu/Edetc/IPSE/educators/activities/supplements/cuttingNano-Handout.pdf.

- For more activities to do from MRSEC, go to Web site: http://mrsec.wisc.edu/Edetc/IPSE/educators/cuttingNano.html

READING MATERIAL

Brezina, Corona. *Careers in Nanotechnology*. New York: Rosen Publishing Group, 2007.
Fritz, Sandy. *Understanding Nanotechnology: From the Editors of Scientific American*. New York: Warner Books, 2002.
Johnson, Rebecca, L. *Nanotechnology (Cool Science)*. Minneapolis, MN: Lerner Publications, 2005.

Poole, Charles P., and Frank J. Owens. *Introduction to Nanotechnology.* Hoboken, NJ: John Wiley & Sons, Wiley-Interscience, 2003.

Ratner, Mark, and Daniel Ratner. *Nanotechnology: A Gentle Introduction to the Next Big Idea.* Upper Saddle River, NJ: Prentice Hall, 2003.

Shelley, Toby. *Nanotechnology: New Promises, New Dangers (Global Issues).* London: Zed Books, 2006.

VIDEOS

What Is Nanotechnology? An engineer from University of Wisconsin-Madison, Wendy Crone is on a mission. She and her interns are creating user-friendly exhibits to teach the public about the nanoworld. http://www.sciencedaily.com/videos/2006-06-11/

Work Force Preparation. Are We Prepared to Get into the Nanotechnology Workforce? Professor Wendy Crone. Conversations in Science. Madison Metropolitan School District. UW-Madison Interdisciplinary Education Group. http://mrsec.wisc.edu/Edetc/cineplex/MMSD/prepared.html

When Things Get Small. Google Video. Describes how small is a nanometer? The film shows how scientists layer atoms to form nanodots. http://video.google.com/videoplay?docid=-2157292956133330853

Careers in Nanotechnology Information Video. Penn State University. http://www.cneu.psu.edu/nePublications.html

WEB SITES

The Foresight Institute: Nonprofit institute focused on nanotechnology, the coming ability to build materials and products with atomic precision, and systems to aid knowledge exchange and critical discussion, thus improving public and private policy decisions. http://www.foresight.org/

National Nanotechnology Initiative: The National Nanotechnology Initiative (NNI) is a federal R&D program established to coordinate the multiagency efforts in nanoscale science, engineering, and technology. http://www.nano.gov/

Scientific American: Nanotechnology articles, some free and some archived, http://www.sciam.com/nanotech/

SOMETHING TO DO

Make a nanoruler to measure objects at the nanoscale. Go to the Lawrence Hall of Science site for the instructions, http://www.nanozone.org/nanoruler_print.htm

2

The Science of Nanotechnology

Nanotechnology is having the ability and the knowledge to manufacture products at the nanoscale—one billionth of a meter. All materials that are one-billionth of a meter are invisible to the human eye. The naked eye can see to about 20 microns. A micron (10^{-6}) is one-millionth of a meter. Even using an optical microscope that uses light energy, we can only see materials that are about 1 micron in size.

In Chapter 1 you read about a number of nanotechnology products that are sold today. These products included pants that are stain-resistant, sunscreens that protect the skin against sun rays, medical dressings that are used for burn victims and special nanoparticles used to make tennis rackets and hockey sticks. So, how can you "see," manipulate, and make products at the nanometer scale, the size of atoms and molecules?

To understand how nanotechnology can be used to manufacture these nanoproducts, you need to review the science of:

- matter and its forms,
- matter's smallest particles—atoms and molecules,
- chemical bonding of atoms and molecules,
- molecular self-assembly, and
- how to manipulate matter at a scale of 1 to 100 nanometers.

WHAT IS MATTER?

Forms of Matter

Look around you. Everything around you is made up of matter. Matter is the air you breathe, the water you drink, the food you eat, the clothes you wear, and the home you live in.

Table 2.1 International System of Measurement [*Système International d'Unités* (SI Units)]

Unit	Abbreviation	Description	Scientific Notation
meter	m	approximately 3 feet	1 m
centimeter	cm	1/100 of a meter, about ½ inch	10^{-2}
millimeter	mm	1/1000 of a meter	10^{-3}
micrometer	μm	1/1,000,000 of a meter, often called a micron	10^{-6}
nanometer	nm	1/1,000,000,000, the size of a single molecule	10^{-9}

Matter is anything that has mass and volume. Mass is the amount of matter in an object and volume is the amount of space occupied by an object. Matter exists in three principal forms, called phases—gas, liquid, and solid.

Gas, or vapor, is the most energetic phase of matter. In gases, the particles (individual atoms or molecules) are far apart from each other and can move about freely. Air neither holds its shape nor its volume because particles move freely through open space. As a result, gases expand to fill the shape of their container. A gas, such as oxygen, is in constant motion and takes the shape of and completely fills any container holding it.

In liquids, the particles are much closer together. So liquids are far more difficult to compress. The particles that make up liquids move about, enabling liquids to change shape easily. A liquid takes the shape of the container holding it.

In solids, the forces between the particles are strong enough to hold the particles together in specific positions causing solids to maintain their shape. As an example, copper holds it shape because the particles stay bound together in a regular pattern.

> What is Matter?. If you have time, you may want to watch the video, Go to: http://www.wpsu.org/nano/lessonplan_detail.php?lp_id=21

Properties of Matter

The different forms of matter have physical properties and chemical properties that are characteristic of individual substances. Physical properties of a substance include color, odor, hardness, density, and the ability to conduct electricity and heat. We distinguish different types of matter by the differences in such physical properties. As an example, the formation of ice is a physical phase change, freezing of water. Each substance has unique phase change properties, such as the temperatures, at which freezing or boiling of the substance occurs, known as the "freezing point" and "boiling point."

Chemical properties describe the ability of a substance to change from one kind of substance into a new and different brand-new substance. The formation of iron oxide (rust) is a chemical change. The more properties you can identify in a gas, liquid, or solid, the better you know the nature of that substance.

PROPERTIES OF MATTER CHANGE AT THE NANOSCALE

The properties of matter depend in part on size. The physical, chemical, and biological properties of matter generally differ at the nanoscale when compared to the larger quantities of the same material. This is due, in part, to the difference in surface area per unit of volume at the nanoscale.

Suppose we have a cube of salt. Its volume area is length times height times depth. Let's say that the block is 1 mm × 1 mm × 1 mm, which is 1 cubic millimeter altogether. (We're using millimeters (mm) instead of nanometers for simplicity.)

The cube has 6 square sides, each of which is (1 mm × 1 mm) 1 square mm in area, so the whole cube has an area of 6 square mm. The ratio of the area (A) of the cube to the volume (V) of the cube is A divided by V (written "A/V") or a ratio of 6/1. Now, in contrast, suppose we have a bigger cube of salt whose edges are 2 mm long instead of 1 mm, so this cube has a volume (2 × 2 × 2) of 8 cubic mm. The area of each side of this salt cube is 2 mm × 2 mm, or 4 square mm, and the area of the cube's entire surface then is 6 × 4 square mm, or 24 square mm. So the area-to-volume ratio of the big block is 24 divided by 8, which equals an A/V ratio of 3/1. So, the bigger cube has a smaller A/V ratio than the smaller cube. This example demonstrates how smaller cubes will have greater ratios of surface area to volume. Table 2.2 summarizes this principle.

This table illustrates how as the sides of a cube get shorter, the volume of a cube gets smaller more quickly than its surface area, so the ratio A/V increases. The ratio A/V is twice as great for the cube of 1 mm as for the cube of 2 mm, and three times as great as the cube of 3 mm. This phenomenon happens on the nanoscale very significantly.

Table 2.2 Surface Area Per Unit of Volume

Length of Side of Cube	Volume of Cube (V)	Surface Area of Cube (A)	Ratio of A/V
3 mm	27	54	2
2 mm	8	24	3
1 mm	1	6	6

Notice how the ratio of area to volume changes as the cube becomes smaller.

For a given substance, increasing the number of nanoscale particles also increases the proportion of atoms on the surface compared with the number of internal atoms. Atoms at the surface often behave differently from those located in the interior. Atoms at the surface have a higher energy state, which means they are more likely to react with particles of neighboring substances. The result is that chemical reactions can take place between atoms and molecules at surfaces acting as miniature chemical reactors.

By modifying materials at the nanoscale other properties such as magnetism, hardness, and electrical and heat conductivity can be changed substantially. These changes arise from confining electrons in nanometer-sized structures. As one example, electrons (subatomic particles) do not flow in streams as they do in ordinary electrical wires. At the nanoscale, electrons act like waves. When electrons act as waves, they can pass through insulation that blocks flowing electrons.

Another example of changes at the nanoscale is that some nano substances that should not dissolve in a liquid actually do dissolve. Some elements like gold also show change at the nanoscale. At the macro scale, gold metal is shiny yellow as you noticed in jewelry. If you break up the gold, to particle 100 nanometers wide, it still looks shiny yellow. But, when you break down the particle of gold to 30 nanometers across, the gold appears bright red. As the particle of gold gets even smaller than 30 nanometers, it looks purple and when it is a little bit smaller it will appear brownish in color. Color can change in some other substances, like gold, at the nanoscale. Physical properties including strength, crystal shape, solubility, thermal and electrical conductivity, and magnetic and electronic properties also change as the size decreases.

Let's look at another example of how a physical property changes at the nanometer scale. At the macro scale, a sheet of aluminum is harmless. However, when the particles of aluminum are cut down in size to 20 to 30 nanometers, the metal can explode.

Volume to Surface Area

Nanomaterials have a large proportion of surface atoms, and the surface of any material is where reactions happen. Because of nanoparticles' huge surface area and thus very high surface activity, nanotechnologists can potentially use much less material. The amount of surface area also allows a fast reaction with less time. Therefore, many properties can be altered at the nanoscale. That's the power of nanotechnology.

> **Did you know?**
> Quantum mechanics is the study of matter and radiation at the atomic level.

MATTER'S SMALLEST PARTICLES: MATTER IS MADE UP OF ELEMENTS

Nanotechnology's raw materials are the chemical elements. Any sample of matter, whether it is a gas, liquid, or solid, is either an element or is composed of several elements. Elements are known as the building blocks of matter. An element consists purely of atoms of one kind. It is a substance composed of pure chemicals that cannot be separated into simpler substances. Therefore, an element is always an element, even at the nanoscale.

Sulfur, copper, carbon, oxygen, iron are some common examples of pure chemical elements. Just four of these—carbon (C), oxygen (O), hydrogen (H), and nitrogen (N)—make up approximately 96 percent of the human body. Much of Earth, including the universe, consists of several elements, such as carbon, hydrogen, oxygen, and nitrogen that are found in our bodies, too. Other common elements include iron, copper, calcium, nickel, potassium, and mercury. Many of these elements are listed on the side of vitamin bottles.

The Periodic Table of Elements

The chemical elements are arranged in a chart called the Periodic Table of Elements. The table of elements represents the building blocks of everything—both living and nonliving. About 90 elements have been identified as occurring naturally on Earth. The Periodic Table of Elements is an important tool for scientists.

The table helps them understand the physical and chemical properties of individual elements.

Table 2.3 The Major Elements Found in the Human Body

Element	Symbol	Approximate Percent by Mass in Human Body
Oxygen	O	65,0
Carbon	C	18.5
Hydrogen	H	9.5
Nitrogen	N	3.3
Calcium	C	1.5
Phosphorus	P	1.0
Potassium	K	0.4
Sulfur	S	0.3
Sodium	Na	0.2
Chlorine	Cl	0.2
Magnesium	Mg	0.1

The Periodic Table can be divided into three basic groups of elements. The groups include metals, nonmetals, and semimetals or metalloids. Semi-metals have the properties of both metals and nonmetals. Chemists use one or two letter symbols to represent elements. As an example, the symbol for aluminum is Al. The symbol for oxygen is O.

- The metal elements make up the majority of the groups of elements—more than 70 percent. Some of these metals include potassium, calcium, sodium, barium, and the more familiar ones include such metals as tin, lead, aluminum, copper, mercury, silver, and gold. Mercury is the only metallic element that is liquid at room temperature. Gold and silver nanoparticles are used in medical research and for treating burns and wounds and in cancer research. Most metals are good conductors of electricity and heat and are malleable—easy to bend and shape. However, electrical conductivity in metal decreases with increasing temperature, whereas in semiconductors, electrical conductivity increases with increasing temperature.
- Some of the nonmetals include chlorine, bromine, iodine, and fluorine. The nonmetals are dull in color, usually brittle, and are poor conductors of electricity and heat. You are not going to find too many of these elements in nanofabrication.
- A few elements are classified as metalloids or semimetals because they are between a metal and a nonmetal. The metalloids include silicon (S), boron (B), germanium (Ge), arsenic (As), and tin (Sn). Metalloids have properties of both metals and nonmetals.

Some of the metalloids, such as silicon and germanium, are used as semiconductors in transistors. Transistors are the fundamental building blocks of a microchip and are built on a silicon base.

Semiconductors

A semiconductor is a common term used in the electronics and the computer field. Semiconductors are generally made from certain elements such as silicon, germanium, and chemical compounds such as lead sulfide.

Semiconductors have a special atomic structure that allows their conductivity properties (both good and not so good) to be controlled by energy from electric currents, electromagnetic fields, or even light sources. As an example, when you apply heat energy to a semiconductor you can increase its conductivity of electricity. These semiconductor properties make it possible to use them to make products such as transistors, integrated circuits, and many other types of electronic devices.

These 10 elements make up most of the matter in the universe.

Element 1: Hydrogen
Element 2: Helium
Element 3: Lithium
Element 4: Beryllium
Element 5: Boron
Element 6: Carbon
Element 7: Nitrogen
Element 8: Oxygen
Element 9: Fluorine
Element 10: Neon

SMALLEST PART OF AN ELEMENT: THE ATOM

The smallest quantity of an element, which has all of the properties of the element, is an atom. An atom is a microscopic structure found in all ordinary matter around us.

Atoms are extremely small, on the order of one to several nanometers wide. A nanoparticle would be a collection of tens to thousands of atoms measuring about 1 to 100 nanometers in diameter. As an example, it would take more than a million atoms to match the thickness of this book page.

Writing with Atoms

In 1989, scientists at the IBM Research Center in San Jose, California, conducted an interesting activity using atoms. The scientists manipulated 35 atoms of the gas xenon (Xe), to write the letters, *IBM. IBM* is the name of their company. The letters were 500,000 times smaller than the type size of letters used in printing this book. To place the atoms in the form of the three letters, the scientists used a special tip on scanning tunneling microscope (STM) to push the atoms into place. The bumps you see in the photo (p. 37) are individual atoms. They have been moved precisely into position, in a row, one half nanometer from each other.

> **Did you know?**
> Xenon (Xe) is a gas most widely used in light-emitting devices called Xenon flash lamps, which are used in photographic flashes, stroboscopic lamps, to excite the active medium in lasers which then generate coherent light, in bactericidal lamps (rarely), and in certain dermatological uses.

See Chapter 3 for more information about nanotechnology microscopes.

INSIDE THE ATOM: SUBATOMIC PARTICLES

The three basic kinds of subparticles that make up the atom consist of neutrons, protons, and electrons.

Neutrons and Protons

Almost all the mass of an atom is in the nucleus. The nucleus lies at the center of the atom. The nucleus consists of two types of particles, protons and neutrons. These two particles are called nucleons, the building blocks of the atomic nuclei.

The protons have a positive electric charge. Neutrons do not have any electric charge; therefore they are an electrically neutral subatomic particles of the nucleus. The strong nuclear force binds protons and neutrons together to make the nucleus.

Electrons

Electrons exist outside the nucleus. Electrons carry a negative charge, equal in magnitude and opposite to the charge of a proton. An electron is 1/1836 the mass of a proton and is much smaller. Electrons carry electrical current and the manipulation of these electrons allows electronic

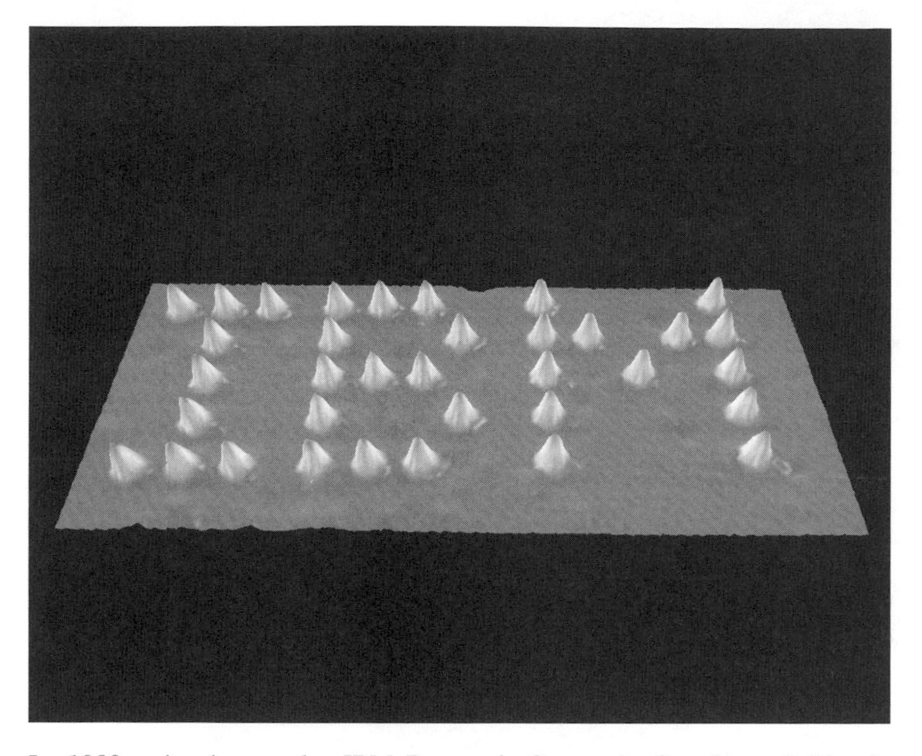

In 1989, scientists at the IBM Research Center in San Jose, California conducted an interesting activity using atoms. The scientists manipulated 35 atoms of the gas, xenon (Xe), to write the letters, IBM. (*Courtesy IBM*)

devices to function. The electron is a subatomic particle that carries an electric charge. When electrons are in motion, they produce an electric current and a magnetic field. Since the nucleus (protons and neutrons) and electrons have opposite charges, electrical forces hold the atom together.

Isotopes

Isotopes of a particular element have the same number of protons, but have a different number of neutrons. If you could change the number of neutrons an atom has, you could make an isotope of that element.

MODELS OF THE ATOMS

Scientists use models to describe and represent the atom. There are several models of the atom to form a concrete or mental picture of what an atom looks like. One model developed by Niels Bohr in 1913 illustrates the atom as having a central nucleus with electrons orbiting

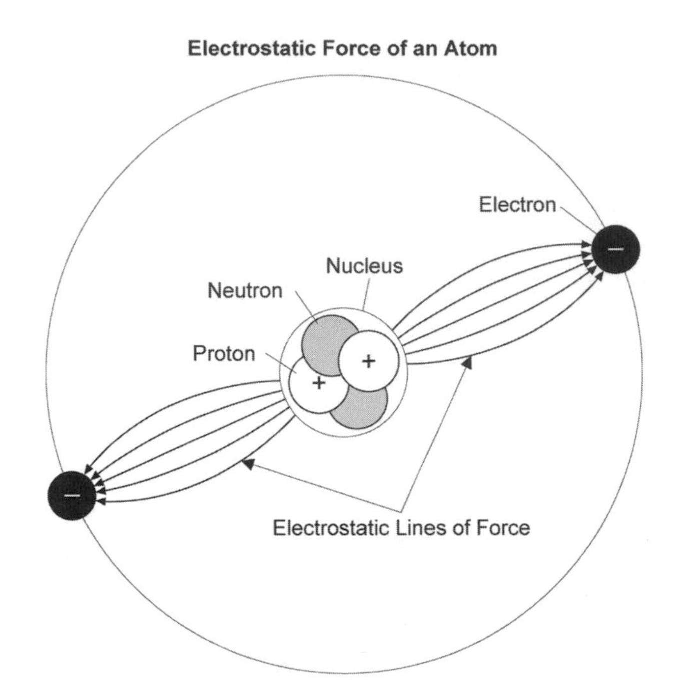

Figure 2.1 Atom diagram. (*Courtesy of Jeff Dixon*)

around it. In the 1920s scientists improved on the model of the atom. One theory is that the electrons move about in a cloud-like region surrounding the nucleus. The electron cloud represents some probable locations of where the electrons are likely to be at a set time.

These models should not be interpreted as any visual representation of an atomic model or as a recreation of an actual atom. The best models are those that are purely mathematical. However, the atomic model images do help as a guide in understanding concepts in chemistry and other sciences.

Early Atomic Theory by Empedocles and Democritus

Empedocles was a Greek philosopher and scientist who lived in Sicily between 492 B.C. and 432 B.C. One of his theories was to describe things around us. Empedocles proposed that all matter was composed of four elements: fire, air, water, and Earth. As an example, stone was believed to contain a higher amount of Earth than the other elements. The rabbit was believed to contain a higher amount of both water and fire than the other two elements.

Another Greek philosopher, Democritus, who lived from 460 B.C. to 370 B.C., developed another theory of matter. He believed that all matter was made of tiny parts. He demonstrated his idea by stating if someone continued to cut an object into smaller and smaller pieces, you would reach the point where the piece was so tiny that it could no longer be cut up and divided. Democritus called these small pieces of matter *atomos,* meaning "indivisible." He suggested that atomos, or atoms, were eternal and could not be destroyed. During his time, Democritus did not have the means that could prove or disprove his theory or even test it.

Atomic Number and Atomic Mass

The number of protons in the nucleus of an atom determines an element's atomic number. In other words, each element has a unique number that identifies how many protons are in one atom of that element. For example, all hydrogen atoms, and only hydrogen atoms, contain one proton and have an atomic number of 1. All carbon atoms, and only carbon atoms, contain six protons and have an atomic number of 6. Oxygen atoms contain 8 protons and have an atomic number of 8. The atomic number of an element never changes.

Atomic masses of elements, however, depend on the number of neutrons as well as protons in an atom and differ from their atomic numbers. Atomic mass is indicated in the periodic table of elements as Atomic Mass Units (AMU) and is a larger value than the atomic number.

ATOMS AND MOLECULES

Nanotechnology is all about the manipulation of atoms and molecules. Atoms seldom exist in a solitary state. Occasionally, atoms join together with their own kind of element. Sometimes, the atom joins together with other elements. When an atom of one element chemically combines with an atom of another kind of element, they form a compound. For example, water (H_20) is a compound made from two elements, hydrogen (H) and oxygen (O). The smallest part of the compound is a molecule, a combination of two or more atoms.

You may want to see the video, *NanoManipulator: Seeing and Touching Molecules.* http://www.nanotech-now.com/multimedia.htm

Molecules range in size from a few atoms to thousands of atoms. How do atoms join together to form molecules?

MOLECULES AND CHEMICAL BONDING

Knowing how molecules bond together is important in understanding how nanostructures can be manufactured.

Molecules are combinations of atoms held together by chemical bonds. A chemical bond is actually an electrostatic force between electrons and protons and between atoms and molecules. One research article stated the electrostatic force is quite strong. Electrostatic forces are more than 1000 times as strong as gravitational forces. The chemical and physical properties of matter result from the ability of atoms to form bonds from electrostatic forces. There are two main types of chemical bonds that hold those atoms together. They are called ionic bonds and covalent bonds.

Ionic Bonding

Atoms can lose or gain electrons forming an ion, which has a net electric charge. A negative charged ion, called an anion (pronounced /AN-ion/), is formed when an atom or a special group of atoms gains additional electrons. When an atom loses some electrons, a positive charged ion is created called a cation (pronounced CAT-ion).

Ionic bonding takes place between two oppositely charged ions, an anion and a cation. Ions may consist of a single atom or multiple atoms, in which a group of atoms is called a "polyatomic ion." Examples of polyatomic anions include: carbonate ion, which is composed of carbon and oxygen; and sulfate ion, which is composed of sulfur and oxygen. An example of a polyatomic cation is ammonium ion, which consists of nitrogen and hydrogen. Cations are usually metal atoms and anions are either nonmetals or polyatomic ions (ions with more than one atom). The attraction of the two charges holds the atoms or molecules together. Electrostatic forces hold ionic bonds together. Many compounds made up of an anion and cation will dissolve in water, so that the two ions separate in the water and make an ionic solution. Ionic solutions are used for treating and cleaning the eye.

Ionic bonding occurs between an element that is a metal and one that is a nonmetal. The majority of geological materials, such as minerals and rocks, feature ionic bonding, predominantly.

Covalent Bonds and Monomers

Covalent chemical bonds are different from ionic bonding. Covalent bonding involves the sharing of a pair of electrons by two atoms. A molecule of water, written H_20, is held together by covalent bonds. The two elements, hydrogen and oxygen, share electrons.

Covalent bonding occurs between two elements that are nonmetals. Much of the matter of substances in your everyday life, the solids, liquids, and gases, are the results of covalent bonding. Covalent bonds occur in all sorts of organic substances, including materials associated with living things, such as plant or animal tissue, food, and the like. Covalent bonding predominates in biological systems, although ions have very important roles in these systems, too.

A Monomer

A small molecule that is held together by covalent bonding is called a monomer (from Greek mono "one" and meros "part"). Examples of monomers are hydrocarbons that consist only of the elements carbon (C) and hydrogen (H). Hydrocarbons are combustible when they combine with oxygen. They are the main components of fossil fuels, which include petroleum, coal, and natural gas. When the covalent bonds of these types of substances are broken they will release energy, which is what occurs when fossil fuels burn.

From Monomers to Polymers

Monomers may become chemically bonded to several other similar monomers to form a polymer. A polymer is a term used to describe a very long molecule consisting of structural units and repeating units connected by covalent chemical bonds. When monomers are linked together to each other during a chemical reaction, it is called polymerization. Often polymerization occurs as a chain reaction, which will continue until a large number of the monomers have combined (polymerized) with each other. The result of polymerization is a chain or other network of linked monomers that can be formed into fibers, sheets, fabrics, foams, or other structures, depending on the type of polymer.

Polymers can occur naturally, mainly in living organisms, or synthetically, through manufacturing. A few general examples of naturally occurring polymers include proteins, starches, and cellulose. Starches are polymers composed of sugars. Latex, from the sap of rubber trees, can be made to polymerize into latex rubber. The genetic materials DNA and RNA found in chromosomes and cells are also types of polymers. Several natural fibers such as wool, silk, and spider web thread are polymers. Scientists and engineers study the formation, structure, and properties of such natural polymers as models for synthetic nanomaterials.

Figure 2.2 Polymer. (*Courtesy of Jeff Dixon*)

Polymers and Nanotechnology

Polymerization is commonly used for making nanoscale materials and other materials as well. Polymerization is used to create the plastic materials found in thousands of strong, flexible, and lightweight products in your everyday life. Some familiar examples of polymers include the common plastics, polystyrene, and polyethylene. These kinds of polymers are used in everything from food containers and packaging materials to cars, boats, and computers. Polyvinyl alcohol is a main ingredient in latex paints, hair sprays, shampoos, and glues, and waxes and oils. Let's look at some examples of applications of polymerization in nanotechnology.

> **Did you know?**
> DNA is another example of a polymer.

Polymer-Based Nanosponges

At the Los Alamos National Laboratory in Los Alamos, New Mexico, scientists have developed a reusable polymer-based nanosponge. The nanosponge has nanometer-sized pores that can absorb and trap organic contaminants in water. The Nanosponge polymer can be used to clean up organic explosives and oil or organic chemical spills especially in water, while decreasing clean up costs associated with present technologies. The nanosponge is made up of polymeric building blocks that form cylindrical cages to trap organics. After the sponges are saturated with contaminants, they can be rinsed with ethanol to remove the contaminants and the nanosponge can then be reused.

The polymer sponge has multiple applications. For example, a polymer designed as a membrane can be placed on a water faucet. The membrane can be used to treat and purify the water for drinking and cooking uses. One advantage of using polymers is that they are inexpensive to manufacture and can be used in a variety of water treatment systems.

Polymer Solar Cells

In the hope of making solar energy more useful and affordable, several scientists have been working on creating organic photovoltaic cells. Their goal is to replace the usual silicon (S) with readily available materials such as carbon (C). If they succeed, designers could one day integrate solar cells into everyday gadgets like iPods and cell phones. Even the energy absorbed by window tinting could be used to power a laptop, for example.

One team of researchers at the University of California have developed an organic photovoltaic cell that uses a polymer, or plastic, material in a unique way. Like other organic solar cells, the plastic material in their prototype includes a polymer material. The material, composed of common chemicals, is sandwiched between conductive electrodes. Photons in the sunlight "knock" electrons from the polymer onto one of the electrodes. This causes an electrical imbalance where one electrode becomes positively charged while the other is negatively charged. When this happens an electrical current is created. See Chapter 8 for more information about solar cells.

A Unique Class of Synthetic Polymers, Dendrimers

One unique class of polymers is dendrimers. Dendrimers play an important role in nanotechnology. Dendrimers, which are synthesized from monomers, are treelike highly branched polymer molecules that are about 3 nanometers across.

Several different kinds of dendrimers have been synthesized. They can be very useful in low-cost plastics such as radiator hoses, weather stripping, and gasket seals. In biotechnology applications, dendrimers are useful for their antiviral properties.

> **Did you know?**
> Dendrimer comes from the Greek word *dendra*, meaning tree. The term dendritic describes anything that has a branched, tree-like structure.

Dendrimers and Drug Delivery

Nanotechnology has the potential to be very useful for improving drug delivery in treating patients with illnesses and diseases. The delivery of drugs into the body cannot be a hit or miss effort. The drug molecules have to find the exact place in the body where they can be the most effective. An anticancer drug needs to be delivered to a tumor where it can do the most good. You do not want the anticancer drug delivered to another organ in the body.

Dendrimers have several advantages in drug delivery; Dendrimers can hold a drug's molecules in their structure and deliver the drug to a particular part of the body such as a tumor. Dendrimers can enter cells easily and release drugs on target and don't trigger immune system responses.

Dendrimers can also be used for chemical analysis and diagnosis of the human body and to locate, diagnose, and then treat tumors or other sick cells. However, more research in safety issues needs to be done.

Dendrimers are of particular interest for cancer applications. One important factor is that it is easy to attach a large variety of other molecules to the surface of a dendrimer. These molecules could include tumor-targeting agents that can pinpoint tumors, deliver an anticancer drug to a tumor, and may detect if the cancer drug is working.

Did you know?
The term *dendrimer* was first used and patented by Donald Tomalia and others and Dow Chemical in the 1980s.

MOLECULAR SELF-ASSEMBLY AND NANOFABRICATION

The term used for the manufacturing of nanoproducts and nanostructures is nanofabrication. One branch of nanofabrication is self-assembly. Self-assembly is a strategy in which objects and devices, such as atoms and molecules, can arrange themselves into an orderly structure or final product without any outside assistance. Self-assembly could occur if you could shake a box of separated puzzle pieces, then look inside the box to see a finished puzzle. The puzzle pieces self-assembled. Materials that self-assemble include snowflakes, salt crystals, and soap bubbles. Each one arranges itself into a pattern.

In the human body, the self-assembly process is similar to the way your bones grow. Individual molecules are formed layer by layer on a surface of the bone. Self-assembly occurs spontaneously when the human body converts food, water, and air into a variety of acids, sugars, and minerals. From these materials cells, blood, tissues, and muscles are created. All of these biological functions are due to the self-assembly of molecules in the human body which continue to build and repair cells and store energy all day long.

What is the advantage of using self-assembling in nanofabrication? One major reason is that to design and assemble nanoparticles atom by atom to make a lot of products is too slow and costly. Also, the atoms are too small for anyone to direct, find, and place individually and quickly.

Self-Assembly. Many researchers are exploring self-assembly of molecules at the nanoscale to produce structures. In this activity, middle school students are modeling how self-assembly takes place. In this activity several LEGO® blocks are placed peg-side down on water. In time, the bricks will interact with the surface tension of the water and with each other to assemble into a pattern. (*Courtesy Dean Campbell*)

SOAP BUBBLES SELF-ASSEMBLE

Soap molecules can self-assemble into bubbles when you blow air through a soapy loop. The soap bubble molecules form two layers that contain a layer of water in between. The soap forms a monolayer on the inside and a monolayer on the outside of the water. Each layer of soap is a self-assembled monolayer. The distance between the outer and inner layers of the bubble is about 150 nanometers. But the soap layer can become a bubble only when the conditions are just right. You need mostly air and very little water.

USING THE SELF-ASSEMBLY STRATEGY TO MAKE PRODUCTS

About one-quarter of the 2,000 or so nanotechnology projects the National Science Foundation now sponsors involve molecular self-assembly.

Molecular self-assembly is the most important of the nanoscale fabrication techniques because it's fast, inexpensive, and produces less waste.

Some of the major companies investing heavily in self-assembly include Merck, Pfizer, 3M, IBM, and Hewlett-Packard. About one-quarter of the 2,000 or so nanotechnology projects the National Science Foundation now sponsors involve self-assembly.

Many soft or fluid consumer products such as foods, paints, detergents, personal care products, and cosmetics contain nanometer structures that are formed by the self-assembly of molecules.

OTHER APPLICATIONS OF MOLECULAR SELF-ASSEMBLY

Self-Assembly in Medicine

Using the self-assembly process in medicine applications could greatly improve orthopedic implants. The average life of an implant today is about 15 years, according to the American Academy of Orthopedic Surgeons. When implants fail in the body, they often break or crack the bone they are attached to. It is also very difficult to get bone to grow on the implant, which is usually made of titanium or ceramics.

Researchers have developed a special coating comprising molecules that self-assemble into a structure similar to that of bone. The research shows that when such a coating is applied to an implant, bone cells adapt to these structures and grow onto them.

In another research project, scientists have designed artificial molecules that self-assemble to form structures that they think could be used to create a piece of artificial spinal cord. The artificial spinal cord could help paralyzed patients regain some mobility. The self-assembled molecule structures would act as a scaffold in which spinal-cord tissue or bone tissue could regenerate.

Self-assembly could change the nature of drug making too. A research group in Rochester, New York, is working on molecules that can self-assemble into drugs. The drugs would have the potential to sense cells with a disease and target them with treatments. The self-assembled molecules bind to specific genes that cause certain genetic diseases and this process blocks the genes from functioning. Of course, self-assembly in drug making still has to be proven nontoxic to humans. Human clinical trials are necessary. There is also the lengthy federal approval process to go through. Plus, scientists have yet to learn how to design the right molecules and figure out how to force them to assemble in certain ways.

NanoSonic, Inc.

Several companies use molecular self-assembly to make products. One of the companies, NanoSonic Inc., is located in Blacksburg, Virginia, The company's team of researchers is headed by the president, Dr. Richard Claus, a professor at Virginia Tech. NanoSonic Inc. makes a patented material called Metal Rubber™ using a molecular layering process, known as electrostatic self-assembly.

Metal Rubber™ looks like brown plastic wrap, and has some amazing properties, including elasticity. "We can stretch it to about 200 to 300 percent of its original length, and it relaxes back," says Dr. Claus. "It's very tough. We can expose it to chemicals. We can put it in jet fuel. We can hit it with acetone. We can boil it in water overnight, and it doesn't mechanically or chemically degrade. We can heat it up to about... 700 Fahrenheit. It won't burn. We can drop it down to about... minus 167 degrees Fahrenheit, and it maintains its properties."

To make Metal Rubber™, the scientists at Nanosonic built it molecule by molecule. The nanotechnology process that is used is called electrostatic molecular self-assembly. "The Metal Rubber™ virtually assembles itself," says Dr. Claus.

The scientists start with a plastic or glass substrate, or base, that they have given an electric charge, either positive or negative. Then they dip the substrate alternately into two water-based solutions. One solution contains ionic molecules that have been given a positive electrical charge (cations). The other solution contains ionic molecules that have a negative charge. If the substrate has a positive charge, it goes into the negative molecules first. The molecules cling to the substrate, forming a layer only one molecule thick. After the next dipping, into positive molecules, a second ultrathin layer forms. Making Metal Rubber™, Dr. Claus explains, is like "making a layer cake."

Dr. Claus says that with Metal Rubber™, nanotechnology has produced a material with many potential uses. One of the most exciting is to make what he calls "morphing aircraft structures. These are aircraft that dynamically change the shape of their wings and their control surfaces during flight," he explains. "Almost the way that a hawk might fly along, see prey, and change its shape to dive down. For a plane, you need a material that's mechanically flexible. But you also need a material with a surface that's controlled by sensors and electrical conductors that allow it to do that sensing and change shape accordingly."

NANO INTERVIEW: DR. RICHARD CLAUS, PRESIDENT OF NANOSONIC

Dr. Richard Claus is president of NanoSonic, Inc., located in Blacksburg, Virginia. Dr. Claus responded to questions from students at the ForwardVIEW Academy in North Kingstown, Rhode Island. The students under the guidance of their chemistry teacher, Ms. Catherine Marcotte, were involved in doing many of the activities in the NanoSonic Kit. The following are questions that they addressed to Dr. Claus.

What made you (Dr. Claus) want to study nanotech?
I was teaching engineering at Virginia Tech and one of the students in my group started a small project in the lab. That was in the early 1990s, and I have been interested ever since.

What college did you go to?
Johns Hopkins University in Baltimore.

What was your major?
As an undergraduate, I majored in Arts and Sciences. As a graduate student, I majored in Engineering.

How did you get started in Nanotechnology?
I was working on the faculty at Virginia Tech and doing work with a number of students in the area of optical fibers. We had a project with a company to put magnetic coatings onto a fiber, and the student mentioned above tried a "nanotechnology" process to apply the coating.

When did you get your first idea about starting NanoSonic?
Again, I was teaching at Virginia Tech, and another company came to us and asked us to do a project with them. The university could not approve the legal terms and conditions of the contract proposed by the company, so we went off campus and started our own company to do the work. Our objective was really just to support a graduate student and not to start a company.

Who comes up with all the NanoSonic ideas?
We all do. We currently have 62 people at NanoSonic, and they are a very interesting mix of individuals with very different technical backgrounds and experiences. Most of our ideas are due to the combination of suggestions from our chemists and engineers and production people. One

of our key laboratory staff people has a background and education in art, and we depend on her extensively for suggestions.

Does nanotechnology have any impact on your life?
Directly, no. But there are lots of parts of my life that relate to nanotechnology that seem more fun now, like why the sky is blue (really) and how I can prevent the inside of my car windows from fogging.

What is the best thing you have ever done involving nanotechnology?
The best thing I have ever done has been hiring good people who know a lot more about many different aspects of nanotechnology and practical manufacturing than I do. They are the ones who do all of the work.

Has studying this field changed the way you look at some things?
Yes, as I mentioned earlier. One example is that nanoparticles are in a number of normal household products, like suntan lotion. A few months ago, a group of us went to the local drugstore and spent about an hour looking at ingredients on the backs of suntan lotion products.

How did you start the company?
When Virginia Tech did not want to pursue the project with the company I mentioned above, we started the company the following afternoon online using a credit card. It took about 15 minutes.

Is your job fun?
Yes, very much.

Can you cut Metal Rubber™?
Yes, with a pair of scissors or a knife. It has the consistency of a mouse pad.

Could you send a small sample of Metal Rubber™ to our school?
Let me see if the people down in the production area have anything we can send.

Could you visit our school?
I would be glad to if I am up your way.

What led to your making Metal Rubber™?
Several of us went to a large international conference and heard a talk by a professor at a major research university. He said that no one would ever make anything big and thick using self-assembly processing. The person next to me, Dr. Jennifer Lalli, turned to me and whispered, "We can do that!" Within a few weeks, Dr. Lalli and her group had made our first Metal Rubber™ materials.

Are you taking the measures to protect the environment from the pollution of the process (is there any?) of making Metal Rubber™ because it will stop a lot of problems in the future?
Yes, we are. The byproducts of the production process are not significant pollutants, but we work closely with our town to properly handle and treat all of the materials that we generate.

Where would you dispose of Metal Rubber™?
In a landfill most probably. Again, it is pretty much like a mousepad.

Is it environmentally safe and biodegradable?
Safe, yes. Biodegradable, no. It is something like milk jug containers—they are safe but will not degrade in a landfill for a long time.

Can you recycle it into something else?
Yes, we can. I was on the phone about this in particular about an hour ago.

How did you hear about nanotech?
Meetings and conferences mostly, and technical journals. I was a professor at Virginia Tech for about 30 years, so I did a lot of reading about science and engineering. I still do.

NANO ACTIVITY: NANOTECHNOLOGY DEMONSTRATION KIT FROM NANOSONIC

You can learn more about self-assembly by doing some of the experiments in a nanotechnology school kit produced by Dr. Claus and the scientists at NanoSonic Inc. The Nanotechnology Demonstration kit provides experiments that are designed for middle and high school science classes. The kit contains all of the materials you need for 24 students. Instructions, worksheets, teacher aids, and a resource CD are provided.

The purpose of the Nanotechnology Demonstration Kit is to allow secondary school students to learn about nanotechnology and next generation materials by making and testing nanostructured materials by themselves using a minimum of laboratory supplies and ordinary tap water.

The kit is divided into five units that will introduce students to the following concepts.

- Nanotechnology and nanostructured materials
- Chemical bonding

- Electrostatics
- Electrostatic self-assembly
- Fabrication of their own nanostructured film

To learn more about the kit, contact NanoSonic, www.nanosonic.com

Ms. Catherine Marcotte, a science teacher and her high school classes, had the opportunity to use the Nanotechnology Demonstration Kit of activities produced by Professor Richard Claus of NanoSonic, Inc, and the manufacturer of Metal Rubber™. Her class interviewed Dr. Richard Claus and asked him questions about his work. I asked Ms. Marcotte to comment on her experiences as a science teacher and her thoughts about the world at the nanoscale. Here are her comments.

"I am currently in my eleventh year teaching varied science topics to small, mixed groups of secondary school students at an alternative public school program called ForwardVIEW Academy in North Kingstown, Rhode Island.

"Most often, each spring or summer, a science theme, such as energy, water, or science, society, and technology as in this year, seems to be in the air that becomes the core of my plans for the coming school year. Other times, the course is chosen and serendipity works its magic. Basic chemistry was the focus that year and, coincidentally, nanotechnology seemed to be on the cover of every magazine.

"Then, I had the pleasure to meet the author Mr. John Mongillo, who brought everything together with a lab kit for high school science involving nanotechnology by NanoSonic, Inc., which he wanted to co-teach with me in my classes. I was thrilled. It fit in perfectly and also provided the perfect segue to this year's technology theme.

"The kit gave students the opportunity to apply the concepts of nanotechnology, such as self-assembly, in making a film on a glass slide one molecule layer at a time! In the unit, students were reminded of atomic structure and learned of electrostatics, and were also made aware of the many careers available in the growing field of nanotechnology.

"Later, through John, the students and I had a rewarding conversation with the President of NanoSonic, Inc., Dr. Richard Claus, by writing questions for Dr. Claus to which he kindly responded by email. The entire experience brought the classes' imaginations far beyond the confines of the classroom, from the uses for flexible Metal Rubber™ to even a lovely house in the misty, morning hills of Blacksburg, Virginia, and the thoughtful musings of a scientist and company president in answer to the question, "Why do you say your job is fun?"

"There is a comfort available in studying nature at its unimaginably small, and learning about the incredible advances in nanotechnology. The rewards are wonder and amazement at the depth of order and beauty in the physical world. While we thinking humans are capable of appreciating this, we are also capable of that which can seem the antithesis of order and beauty. Nature can seem more reliable. Learning more about our world on the nanoscale and realizing our own membership in it can, I think, provide a sense of reassuring continuity."

READING MATERIAL

Gross, Michael. *Travels to the Nanoworld: Miniature Machinery in Nature and Technology.* New York: Perseus Books Group, 2001.

Jones, R. L. *Soft Machines: Nanotechnology and Life.* Oxford, U.K.: Oxford University Press, 2004.

Krummenacker, Markus, and James J. Lewis. J. *Prospects in Nanotechnology: Toward Molecular Manufacturing.* New York: John Wiley & Sons, 1995.

Mulhall, Douglas. *Our Molecular Future: How Nanotechnology, Robotics, Genetics, and Artificial Intelligence will Transform Our World.* Amherst, N.Y: Prometheus Books, 2002.

Sargent, Ted. *The Dance of the Molecules: How Nanotechnology Is Changing Our Lives.* New York: Thunder's Mouth Press, 2006.

Smalley, R.E. *Carbon Nanotubes: Synthesis, Structure, Properties and Applications.* New York: Springer, 2001.

VIDEOS

(Please note that playing the videos will require different plug-in applications, which means you may need to download the proper video player.)

What is Matter? http://www.wpsu.org/nano/lessonplan_detail.php?lp_id=21

What is a Molecule? http://www.wpsu.org/nano/media/Molecule.mov

Electrostatic Self-Assembly. NanoSonic. http://www.nanosonic.com/schoolkits/schoolkitsFS.html

Nanoparticles—Chemistry, Structure and Function. Presented by: Karen L. Wooley, PhD, Professor Washington University in Saint Louis, Department of Chemistry. http://www.blueskybroadcast.com/Client/ARVO/

Dendritic Polymer Adhesives for Corneal Wound Repair, Presented by: Mark W. Grinstaff, PhD, Associate Professor of Biomedical Engineering and Chemistry, Metcalf Center for Science and Engineering. http://www.blueskybroadcast.com/Client/ARVO/

Nanowires and Nanocrystals for Nanotechnology. Yi Cui is an Assistant Professor in the Materials Science and Engineering Department at Stanford. video.google.com/videoplay?docid=6571968052542741458

WEB SITES

Big Picture on NanoScience: http://www.wellcome.ac.uk/node5954.html

Nanozine—This site is a Nanotechnology magazine with many definitions and links to other nanoscience sites.

Foresight Nanotech Institute: http://www.foresight.org/

NASA and Self-Replicating Systems:http://www.zyvex.com/nanotech/selfRepNASA.html

How Stuff Works: How Nanotechnology Will Work:

Animated narrative shows how nanotechnology has the potential to totally change manufacturing, health care and many other areas.

http://www.howstuffworks.com/nanotechnology.htm

SOMETHING TO DO

Make soap bubbles to learn about self-assembly. Exploratorium.

http://www.exploratorium.edu/snacks/soap_bubbles.html

Nanoscale Activity: NanoSugar. Introduces the idea of a nanometer and ties in the surface area to volume concept, University of Wisconsin.

http://mrsec.wisc.edu/Edetc/IPSE/educators/nanoSugar.html

3

The Nanotechnology Tool Box

How do you "see" and manipulate matter at the nanoscale? What kinds of tools do you need to observe what atoms and molecules look like, and once you "see" them, how do you pick them up and move them around?

This chapter will focus on some of the tools, such as the present state-of-the-art microscopes and how they operate in the nanoworld.

> *Amazing Creatures with Nanoscale Features,* the video is an introduction to microscopy and applications of nanoscale properties. http://www.cneu.psu.edu/edToolsActivities.html.

When you have time, you can see a video of one of these state-of the-art microscopes in action.

OPTICAL MICROSCOPES

Many of us have used an optical microscope at home or in the classroom. Optical microscopes focus visible light through a combination of lenses to produce a magnified image of an object, say a flea's leg. Scientists used glass lenses and mirrors to focus and magnify light on an object. To increase the magnification of a microscope, more lenses are added to the instrument.

Optical microscopy can distinguish objects in the micrometer range, which is about 10^{-6} meters. However, the resolution power of these instruments is limited to revealing objects down to about 200 nanometers to 250 nanometers magnifications. The tiniest objects we could see would include red blood cells and small bacteria.

Optical microscopes have technical limitations too. A bacterium can be seen through a microscope that works with visible light because the bacterium is larger than the wavelength of visible light. However, tinier objects such as atoms, molecules, and viruses are invisible because they are smaller than the wavelength of visible light. Therefore, they do not reflect the light of the image toward our eyes.

Although the optical microscope allows us to see many images, it leaves a lot of objects that we cannot see, such as viruses, atoms, molecules, and the DNA helix. To see these objects, we need other kinds of nonoptical microscopes, called nanomicroscopes.

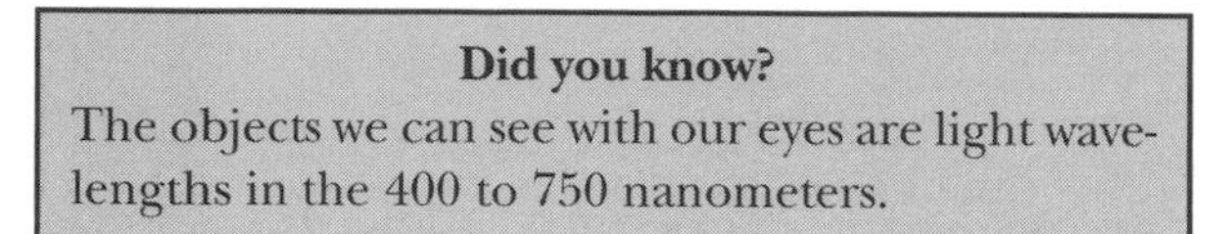
Did you know?
The objects we can see with our eyes are light wavelengths in the 400 to 750 nanometers.

SCANNING PROBE MICROSCOPES

Scanning probe microscopes (SPM) are nano instruments that are used to study the surface of materials at the nanoscale. The two major kinds are the scanning tunneling microscope (STM) and the atomic force microscope (AFM). SPM enables resolution of features down to about 1 nanometer in height, allowing imaging of single atoms under ideal conditions.

Both scanning probe microscopes can provide pictures of atoms on or in surfaces. The study of surfaces is an important part of physics, with particular applications in semiconductor physics and microelectronics. In chemistry, surface chemical reactions also play an important part, for example, in the process of catalysis. A catalyst is a substance that produces a change that is used to increase the rate of a chemical reaction by creating conditions that make the reaction easier to proceed.

SURFACE AREA TO VOLUME AT THE NANOSCALE

Why is the microscope study of surfaces and surface areas so important? Remember in Chapter 2, you read about the difference in surface area per unit of volume at the nanoscale.

In the nanoworld, the relationship between surface area and volume is important because the ratio of surface area to volume increases as objects get smaller and smaller. See illustration and example in Chapter 2. Therefore, since most chemical and physical changes happen on surfaces, the more surface area the object has the more physical or chemical changes that can take place. Confused? The next time you break up one large sugar cube into smaller pieces, you are creating more surface area in each sugar particles per volume. That is why the particles will dissolve faster (a physical change) in a liquid than the one large cube. The sugar particles have more surface area per volume than the one large cube.

Scanning Tunneling Microscopes (STM)

The first scanning probe instrument was the scanning tunneling microscope (STM). The scanning tunneling microscope (STM) can show three-dimensional images of individual atoms at the surface.

Gerd Binnig and Heinrich Rohrer of IBM's Zurich Laboratory in Zurich, Switzerland invented this microscope in 1981. The two scientists received a Nobel Prize in Physics in 1986 for their innovation of the STM.

How Does the STM Work?

Scanning tunneling microscope (STM) uses a probe (the tip) to move over the surface of the material being studied.

The tip of the probe, usually composed of tungsten or platinum, is extremely sharp. The tip, which consists of a single atom, is attached to cantilever that looks like a toy diving board with a sharp tip under one end of it.

The tip slowly scans across the surface at a distance of only an atom's diameter. The tip is raised and lowered in order to keep the signal constant and maintain the distance. A small force measures the attraction or repulsion between the tip and the surface of the sample during scanning. These forces, causing an up and down movement of the cantilever, are monitored by a laser beam reflecting off the cantilever surface. Recording the up and down movement of the tip makes it possible to study the structure of the surface sample. A profile of the wiggled surface is created, and from that a computer-generated image is produced. The final image is a contour map of the surface showing trenches and valleys. The STM works best with conducting materials, but does not work well with insulators such as rubber.

The STM has many uses and applications in nanotechnology. The STM can image materials from DNA samples and other biological molecules. Another example is that STM has been used to study the surface of operating battery electrodes through which electricity enters or leaves.

Did you know?

The Scanning Tunneling Microscope (STM) is very sensitive to vibrations. Just someone sneezing in the same room, can jam the microscope's tip into the sample and damage the experiment.

Atomic Force Microscopes (AFM)

The atomic force microscope is sometimes called a Scanning Force Microscope. The AFM measures the interaction forces between probe tip and sample.

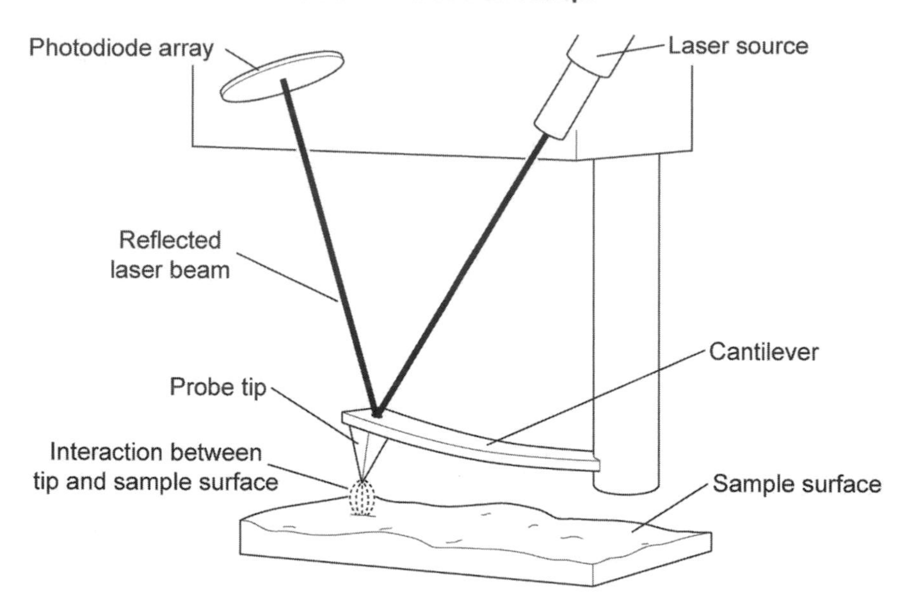

Figure 3.1 How an AFM works. (*Courtesy of Jeff Dixon*)

How Does the AFM Work?

AFM is part of a family of instruments called scanning probe microscopes (SPM). AFM uses a probe moving across the sample's surface to identify its features. The probe is a sharp tip, usually made of silicon, at the end of a cantilever that bends in response to the force between the tip and the sample being viewed. Any deflection from the sample is measured and the AFM records the surface topography. The topography is the projections (lumps), depressions (pits), and textures of a surface. The tip never touches the sample surface because it could damage it. AFMs are widely used to measure surface topography of many types of samples.

The atomic force microscope can be operated under two different conditions (in air or in a vacuum) and two operating modes (contact or noncontact). Whatever the condition or mode, the basic operating principles of the AFM remain the same. The AFM uses a probe that has a tip mounted on a flexible cantilever. The tip is slowly scanned across the surface of a material, just a few angstroms away from the surface (noncontact mode) or in contact with it (contact mode). An angstrom is one tenth (0.1) of a nanometer. The force between the

Table 3.1 Some Samples of Objects That Can Be Observed by an Atomic Force Microscope

Cells: Bacteria, Plants
Ceramics
Fiber: Hair, Nanotubes
Metals
Minerals
Molecules: Self-Assembled Monolayers, Proteins
Paper
Particles: Nanoparticles, Quantum Dots
Plant Surfaces: Leaves, Fruit
Polymers

Note: AFM can image objects that are solid and have a surface.

atoms on the surface of the material and those on the tip cause the tip to deflect. This deflection can be recorded in various ways, the most common of which uses a laser focused on the top of the cantilever and reflected onto photodetectors. A laser is a special type of amplified light beam and is an essential part of CD and DVD players and recorders.

The photodetector signals are used to map the surface topography of samples with resolutions down to the nanoscale. At the nanoscale, the microscope is able to determine roughness, grain size, and features.

The number of applications for AFM has increased dramatically since the technology was invented in 1986. Presently, the AFM microscopes are used for a variety of applications in industrial and scientific research that includes such areas as microelectronics, telecommunications, biological, chemical, automotive, aerospace, and energy industries. Researchers also employ the AFM as a tool to investigate a wide range of materials such as thin and thick film coatings, ceramics, composites, glasses, synthetic and biological membranes, metals, polymers, and semiconductors. The continuing application of the AFM systems will lead to advancements in such areas as drug discovery, life science, electrochemistry, polymer science, biophysics, and nano biotechnology.

Did you know?

An angstrom is very small. An angstrom is a unit of length equal to one-hundredth millionth of a centimeter or 10^{-8}. Remember, a nanometer is 10^{-9}. Chemists and physicists measure the distances in molecules or wavelengths of light in angstroms.

In this activity, students are constructing the cantilever section of a LEGO® model of a scanning probe microscope. LEGO® models are built to understand how a scanning probe microscope works. (*Courtesy Dean Campbell*)

The Advantages of AFM

The difference between AFM and other microscopy techniques is the quality and the measure of resolution. While electron and optical microscopes provide a standard two-dimensional horizontal view of a sample's surface, AFM also provides a vertical view. The resulting images show the topography of a sample's surface. While electron microscopes work in a vacuum, most AFM modes work in dry or liquid environments. AFM does not require any special sample preparation that could damage the sample or prevent its reuse.

AFM also can be used for imaging any conducting or nonconducting surface properties of a substance, while the STM is limited to imaging only conducting surfaces of a sample. The AFM is a major part of the $1 billion nanotechnology measurement tools market that is expected to continue to grow at about 20 percent a year.

AFM Tips

AFM tips are generally made of silicon or silicon nitride. For most applications, pyramid-shaped silicon nitride tips are used. They are relatively durable and present a dry surface to the sample. Conical silicon tips are often used for biomolecular applications because they are very sharp and present a wet surface. However, they are relatively less durable.

Conventional silicon tips are good for measuring relatively flat surfaces, but they do not penetrate the crevices that often exist in small devices and structures. The silicon tips also wear out quickly, reducing image resolution.

Magnetic Force Microscopes

The magnetic force microscope (MFM) is another scanning probe microscope. This microscope has a magnetic tip that is used to probe the magnetic field above the surface of a sample. The magnetic tip, mounted on a small cantilever, measures the magnetic interaction between a sample and a tip. The MFM was developed as an evaluation tool for measuring the magnetism levels in products.

INTERVIEW: NATHAN UNTERMAN

Mr. Nathan A. Unterman is a physics teacher at the Glenbrook North High School in Northbrook, Illinois. He teaches physics to high school juniors and seniors. He and his students have built and used LEGO® models of the atomic force microscope (AFM) and the magnetic force microscope (MFM). The author asked Mr. Unterman about how his interests in nanotechnology and his work in building LEGO® models.

How did you get interested nanotechnology?
I attended a Research Experience for Science Teachers program in the Materials Science department at Northwestern University a few years ago.

Describe some of your experience in developing Lego models of an Atomic Force Microscope and a Magnetic Force Microscope.
I was looking for a physical model that could be used in the classroom for scanning probe microscopes. I also wanted it to be something where quantitative data could be measured. The University of Wisconsin had a prototype for a LEGO® AFM. I then created a method for

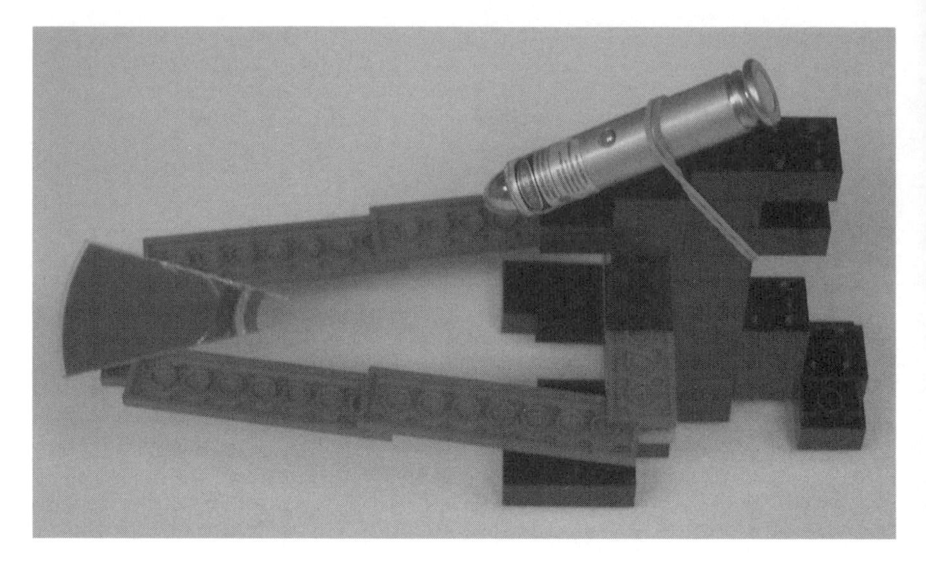

This is a completed LEGO® model of a scanning probe microscope. Professor Dean Campbell and other researchers conceptualize the idea of building a model of a scanning probe microscope from LEGO® building blocks. To construct the model, refer to instructions at the end of this chapter. (*Courtesy Dean Campbell*)

quantitative engagement. This engagement included the use of Excel surface plotting.

How important is the field of nanotechnology in the subject you teach?
In science in general, it is important. In secondary physics, it has not yet worked its way into the curriculum.

What advice would you give students who want to explore a career in nanotechnology?
Learn as much science, math, and writing as you can, and make sure you balance that with the fine and applied arts.

Please describe how the LEGO® model of the atomic force microscope works.
The probe is moved over a tray containing an array of balls that represents the atoms of a sample. The beam from a laser pen in the probe is reflected off the cantilever of the probe as it scans points on the surface under study. The position of the reflected beam is measured on a wall-mounted grid and recorded. These positions are recorded in a spreadsheet and analyzed using the surface plot-graphing tool.

If a teacher or student wanted to construct the LEGO® model of the scanning probe microscope, would you be able to assist them?
They can refer to the article published in the December 2006 *Science Teacher*, or email me at: nunterman@newton.dep.anl.gov

ELECTRON MICROSCOPES

Several different types of electron microscopes exist. Two major ones include the transmission electron microscopy (TEM) and the scanning electron microscopy (SEM). Electron microscopes use electron beams instead of visible light, enabling resolution of features down to a few nanometers.

> Measuring Electrical Properties with an Electron Force Microscope. Professor Wendy Crone, Madison Metropolitan School District. http://mrsec.wisc.edu/Edetc/cineplex/MMSD/scanning3.html

Electron microscopes use a beam of high-energy electrons to probe the sample. Electron microscopes are scientific instruments that use a beam of highly energetic electrons to examine objects on a very fine scale. High quality electron microscopes can cost from $250,000 to $1,000,000. They are one of the most useful instruments in laboratories.

A Scanning Electron Microscope

A scanning electron microscope (SEM) can analyze materials for information about topography, chemical composition, contamination, grain size and thickness with far greater depth of field than is possible with optical microscopy. The depth of field basically means how clear a three-dimensional image looks.

The topography information analyzed by the SEM includes the surface features of an object such as its texture and hardness. A microscope operator can observe the shape and size of the grain particles making up the object and notice if there is any contamination. The composition data includes the elements and compounds that the object is composed of. How the atoms are arranged in the pattern of the object can be detected by the crystallographic. The pattern in the crystal is formed by the way the atoms in a solid material are connected to one another.

ERNST ABBE

In the 1870s, a man named Ernst Abbe explained why the resolution of a microscope is limited. He said that since the microscope uses visible light and visible light has a set range of wavelengths, the microscope can't produce the image of an object that is smaller than the length of the light wave.

The Transmission Electron Microscope

The transmission electron Microscope (TEM) was the first type of electron microscope to be developed. It was developed by Max Knoll and Ernst Ruska in Germany in 1931. The transmission electron microscope (TEM) operates on the same basic principles as the light microscope but uses electrons instead of light. As mentioned earlier, what you can see with a light microscope is limited by the wavelength of light. TEMs use electrons as a "light source" and their much lower wavelength makes it possible to get a resolution a thousand times better than with a light microscope. The enlarged version of the image of the specimen appears on a fluorescent screen or in a layer of photographic film.

You can see objects to the order of .2 nanometers. For example, you can observe and study small details in the cell or other different materials down to near atomic levels. The microscope's high magnification range and resolution has made the TEM a valuable tool in medical, biological, and materials research.

Transmission electron microscopy has had an important impact on the knowledge and understanding of viruses and bacteria. The improvement in resolution provided by electron microscopy has allowed visualization of viruses as the causes of transmissible infectious disease. Researchers are continuing the use of electron microscopy in the investigation of such pathogens as SARS and the human monkeypox virus. SARS is a severe acute respiratory disease in humans which is caused by the SARS virus. Monkeypox is a rare smallpox-like disease that is most common in the rain forests of central and West Africa.

How Does the TEM Work?

A light source at the top of the microscope emits a beam of electrons that travel through a vacuum in the column of the microscope. Instead of glass lenses focusing the light source in the light microscope, the TEM uses electromagnetic lenses to focus the electrons into a very thin beam. The electron beam then travels through the specimen you want to study. Depending on the density of the material present, some of the electrons are scattered and disappear from the beam. At the bottom of the microscope the rest of the electrons hit a fluorescent screen. The screen shows a shadow-like image of the specimen that can be studied directly by the operator or photographed with a camera.

Scanning Electron Microscope

The first Scanning Electron Microscope (SEM) debuted in the 1940s but the first commercial instruments were produced in the 1960s.

The SEM is a type of electron microscope capable of producing high-resolution images of a sample surface. Due to the manner in which the image is created, SEM images have a characteristic three-dimensional appearance and are useful for judging the surface structure of the sample. This scanning electron microscope has a magnification range from15x to 200, 000x and a resolution of 5 nanometers.

The SEM has the ability to image large areas of the specimen that includes thin films and bulky materials as well. In general, SEM images are much easier to interpret than TEM images.

BUGSCOPE

The Bugscope project is an educational outreach program for K-12 classrooms. The project provides a resource to classrooms so that they may remotely operate a scanning electron microscope to image "bugs" or "creatures" at high magnification. The microscope is remotely controlled in real time from a classroom computer over the Internet using a Web browser. BugScope provides a state-of-the-art microscope resource for teachers that can be readily integrated into classroom activities. Students can peek at extreme close-up views of the insect world via their school computer labs and for free.

The BugScope project was developed by the BugScope Project Team and the Imaging Technology Group at the Beckman Institute for Advanced Science and Technology at the University of Illinois at Urbana-Champaign. The scanning electron microscope costs about half a million dollars, and its main function is for research done at the university by graduate and postdoctoral students as well as industry.

The team at the institute sets up preset views of the "creatures" provided by the school. The classroom teacher and students, from a computer station at their own school, can operate the microscope for other views. Students watch the images projected on a screen in the front of the computer lab, and can use their own computer stations to ask questions from the scientists at the center in Urbana.

The classroom has ownership of the project. The students design their own experiments and provide their own bugs to be imaged in the microscope. Bugscope provides resources pages with helpful links related to electron microscopy and bugs. The BugScope Web site: http://bugscope.beckman.uiuc.edu/

Bugscope also offers a virtual microscope activity. They recently developed a Virtual Microscope that allows for Bugscope-type viewing (without an Internet connection) from precaptured high-resolution image datasets. They have built both a Virtual Scanning Electron Microscope (Virtual SEM or VSEM) and a Virtual Light Microscope (VLM). Go to their Web site and download the Virtual Microscope and give it a try for free!

HITACHI TABLETOP MICROSCOPE

Hitachi has developed its first tabletop electron microscope, the TM-1000. The TM-1000 is designed for researchers who want to break through the limits of the light microscopy and close up to sophisticated magnifications up to 10,000.

The microscope is compact, easy to use, is ideal for a wide variety of applications, and has superior resolution for higher magnification than an optical microscope. The TM-1000 provides a real alternative to optical microscopes, stereomicroscopes, and confocal laser scanning microscopes. It has applications for many areas including life science, food, cosmetics, health care, pharmaceutical, textiles, materials science, semiconductor, and education.

NANOFABRICATION CLEANROOM FACILITIES

Scientists need a dust-free workspace to use electron microscopes and other tools effectively to build small-scale circuits and machines. The workspace facility is called a cleanroom and it is an important part of nanotechnology research. In fact it is nearly impossible to conduct nanoscale research without a cleanroom environment.

Why a cleanroom? On the scale of a nanometer, any particle that is 300 nanometers is huge and is capable of causing short circuits in a nanoscale electronic circuit.

Cleanroom environments can cost approximately 3 million dollars to install. Most cleanrooms consist of an air filtration system as well as temperature and humidity control systems. The system constantly moves the air down to the floor where it is sucked back into the return air system and then cleaned and sent back into the room. Some cleanrooms have yellow illumination that allows researchers to work with light sensitive materials. The yellow illumination prevents light-sensitive material from being exposed to ultraviolet light.

As researchers enter the cleanroom, special equipment removes dirt and dust particles from their shoes. The researchers then put on safety glasses, gloves, and coveralls over their street clothes, shoes, and hair. All of the garments are made from lint-free fabrics that are antistatic. It

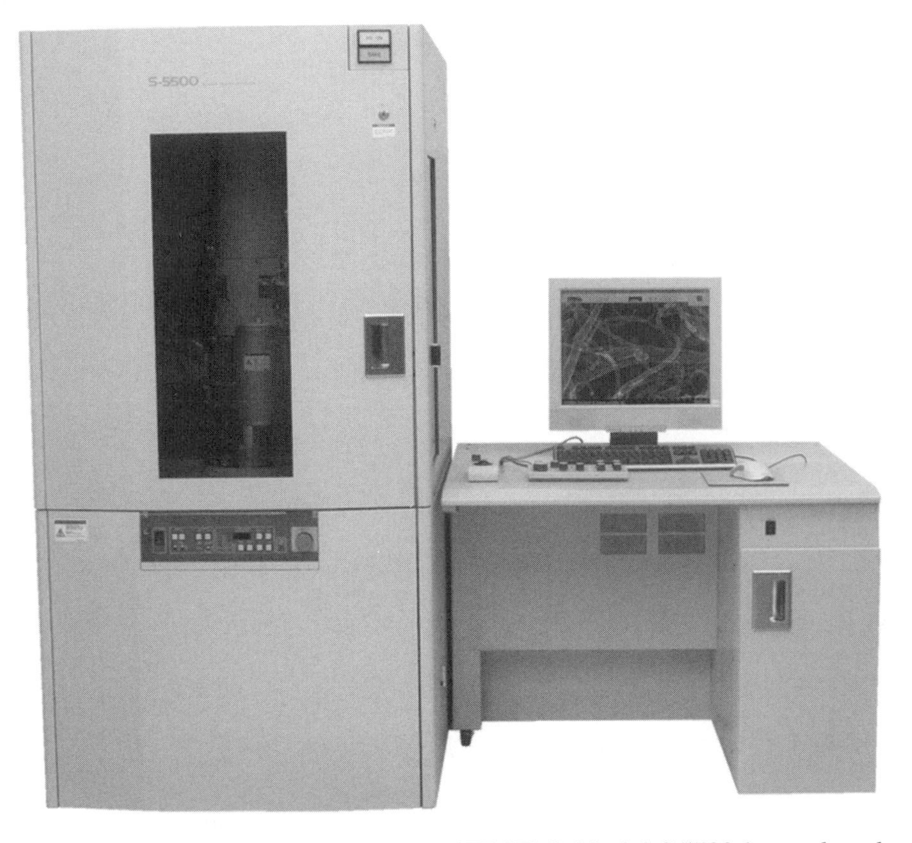

The Scanning Electron Microscope (FE-SEM) Model S-5500 is produced by Hitachi High-Technologies Corporation. The capabilities of the instrument bridge many research fields that include pharmaceutical, biological, and food industry research and nanomanipulation devices for advanced carbon nanotube research. (*Courtesy Hitachi High-Technologies Corporation (Japan)*)

may take 40 steps or more to complete the steps needed to get dressed before entering the cleanroom. Most of us get up in the morning using between 6 and 9 steps to get dressed.

SCANNING ELECTRON MICROSCOPES AND PHOTOLITHOGRAPHY

Today the scanning electron microscope is mainly used for the study of surfaces of materials as well as for the investigation of transparent specimens. Two major applications for the scanning electron microscope are for analyzing and inspecting objects and for photolithography.

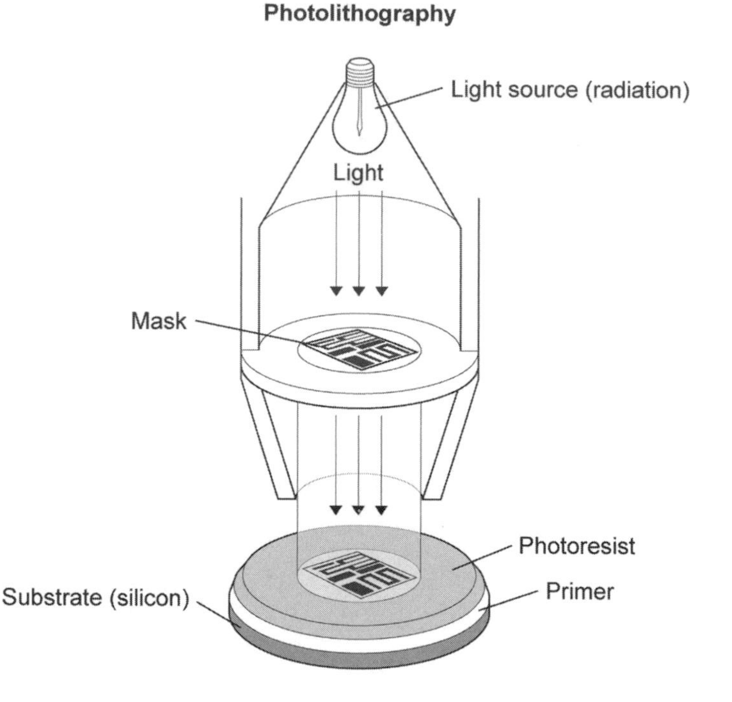

Figure 3.2 Photolithography. (*Courtesy of Jeff Dixon*)

Photolithography is a process used to build components on a micrometer-sized scale. The process uses light to transfer a pattern of say, a microchip part, from a mask to a light-sensitive chemical photoresist that forms an image of the pattern on a substrate. The method is like using silk-screening to make pictures and images from pattern masks on T-shirts. Photolithography has a broad range of industrial applications, including the manufacture of semiconductors, flat-panel displays, micromachines, and disk heads.

NANO-IMPRINT LITHOGRAPHY

Today several companies are researching and experimenting with different kinds of lithographic techniques using nanotechnology. Two techniques include Dip Pen Nanolithography and Thermal Dip Pen Nanolithography.

Dip Pen Nanolithography

Dip Pen Nanolithography (DPN) is a soft-lithography technique that uses an AFM scanning probe tip to draw nanostructures. In the DPN process, a probe tip is coated with liquid ink, which then flows onto the surface to make patterns wherever the tip makes contact. This direct-write technique offers high-resolution patterning capabilities for a number of molecular and biomolecular "inks" on a variety of substrates. Substrates are the base material that the images are printed on.

Some of the applications of the DPN technique include sol gel templates that are used to prepare nanotubes and nanowires, and protein nanoarrays to detect the amount of proteins in biological samples such as blood.

Thermal Dip Pen Nanolithography

Researchers from the Georgia Institute of Technology and the Naval Research Laboratory (NRL) have developed another method for writing nanometer-scale patterns onto a variety of surfaces.

Their writing method is called thermal dip pen nanolithography, which is an extension for dip pen nanolithography. The tDPN method uses easily melted solid inks and special AFM probes. The probes have built-in heaters that allow writing to be turned on and off at will. The tDPN could be used to produce features too small to be formed by light-based lithography. Other applications of the tDPN include using the process as a soldering iron for repairing circuitry on semiconductor chips. The thermal materials also provide sharper features because they do not spread out like liquid inks.

Did you know?

The use of low-temperature, scanning electron microscopy can be used by ski-run operators to determine how effective their artificial snow operations are in lengthening the ski season because of the contrast between artificial and natural snow crystals.

NASA VIRTUAL LAB

How would you like to learn how to use an electron microscope in a virtual lab? Besides the BugScope project, NASA provides a virtual lab online of an electron microscope for schools, pupils, and teachers. The free JAVA program allows users a good sense of what the instrument can do. By downloading the free JAVA program for both PC and Mac, users can pan, zoom, and even use a built-in ruler to measure such items as a beetle's leg, crystals from a human kidney stone, and other objects.

Virtual Lab completely emulates a scanning electron microscope and allows any user to zoom and focus into a variety of built-in microscopic samples. It also comes with a set of educational materials such as a demo on how a SEM works and movies of the real thing in action.

The Virtual Microscope, which includes free downloading, is available on Sourceforge, http://virtual.itg.uiuc.edu/training/

See Chapter 9 for more information about the NASA Virtual Microscope Web site.

NANOMANIPULATORS

One of the major goals in the future is to manufacture products at the nanoscale. In order to achieve that goal, researchers will need to use tools that can grab, pickup, push, pull, turnover, tap down, stack, and move atoms and molecules. To develop the tools needed to do this kind of work will take more research time.

For now, nanotechnololgists are using and are continuing to develop new nanomanipulators—tools to manipulate objects in nanometers. Nanomanipulators are and will be important tools in nanotechnology research and development,

Presently, scanning probe microscopy, such as the atomic force microscope and the scanning tunneling microscope, is one method to manipulate objects in nanometers. As an example, the AFM have been used to move atoms, carbon nanotubes, and to test electronic circuit boards and integrated circuits.

A few companies are providing add-ons or interfaces to scanning probe microscopes such as scanning tunneling microscopes (STM) and atomic force microscopes (AFM). Medical researchers have used a special kind of nanomanipulator to study fibrin fibers, the major components of blood clots. The researchers hope to gain insight into the healing process by observing the strength and mechanical properties of blood clots under a variety of conditions.

One company has a kind of nanomanipulator that offers several positions that grasp, move, and test nanoscale samples. This action allows simultaneous manipulation, imaging, and testing of samples.

NANOFABRICATION

Nanomanipulation will play a big role in nanofabrication. Nanofabrication is the design and manufacture of devices with dimensions measured in nanometers.

Nanofabrication is of interest to computer engineers who are interested in researching and building super high-density microprocessors that can be used in large computer mainframes, microcomputers, and in handheld computers. The goal is to be able to get each data bit stored in a single atom.

Did you know?
The microprocessor or microchip starts up a goes to work when the computer is turned on performs arithmetic and logic operations that spond to and process basic instructions that dı the computer.

If it can be done, a single atom might even be able to represent a byte or word of data.

Nanofabrication has also caught the attention of the medical industry, the military, and the aerospace industry. Nanofabrication methods in the lines of miniaturization (top-down) and self-assembly (bottom-up) are essential for the development of nanotechnology.

Top-down fabrication can be likened to making a baseball bat from a block of wood. The original block of wood is cut down until the desired shape of the bat is achieved. That is, you start at the beginning or the top and slowly work your way down removing any wood in the shaping of the bat that is not needed.

The most common top-down approach to fabrication involves lithographic patterning techniques mentioned earlier in this chapter. The lithographic process uses short-wavelength light sources.

A major advantage of the top-down approach is that the parts are both patterned and built in place, so that no assembly step is needed.

Bottom-up nanofabrication is the opposite of top-down nanofabrication. The bottom-up nanofabrication is to build nanostructures atom by atom using either self-assembly techniques or manipulating atoms by employing scanning probing microscopy.

Now many industries are adapting nanofabrication technology and will use it even more in the future. Some of the companies include those that produce fiber-optic communications, pharmaceuticals, and microelectronics. According to the Federal Government, the unprecedented spread of nanofabrication and nanotechnology is likely to change the way almost everything is manufactured—from vaccines to computers to automobile tires to objects not yet discovered.

NANO INTERVIEW: ASSOCIATE PROFESSOR DEAN CAMPBELL, Ph.D., BRADLEY UNIVERSITY

Dean Campbell, Ph.D., is an associate professor of chemistry at Bradley University in Peoria, Illinois. Dr. Campbell talked with the author of *Nanotechnology 101* about the booklet, *Exploring the Nanoworld with LEGO® Bricks*, scanning probe microscopy, and ideas for building a *LEGO®* model of a scanning probe microscope.

What college did you attend and what was your major field?

I earned my bachelor's degree in chemistry from the University of Wisconsin-Green Bay and my Ph.D. in chemistry from Northwestern University.

Dean Campbell, Ph.D., teaches chemistry at Bradley University in Peoria, Illinois. (*Courtesy Duane Zehr*)

How did you get interested in nanotechnology?
When I was an undergraduate, I was involved in a project studying the surface chemistry of coal ash particles, and I was intrigued by how the chemistry of surfaces could be so different from bulk material. In graduate school I worked with surfaces of a different sort, self-assembled monolayers. These types of monolayers are a rather simple form of nanotechnology, as they tend to have nanoscale thicknesses.

What were your favorite subjects in high school?
I liked a lot of the sciences, because I like to understand how things work. Ironically, because I was slow in chemistry lab, I thought in high school that I would go into any science field EXCEPT chemistry. I changed my mind as an undergraduate because the field was so interesting and employable.

What is scanning probe microscopy (SPM)?
Scanning probe microscopy (SPM) is a method for mapping surface forces of materials on the atomic scale. By mapping these forces, much can be learned about the surfaces of materials, where many interesting and complex phenomena occur. For example, many chemical reactions involving solids are dependent on the nature of their surfaces. Scanning probe microscopy includes the methods of atomic force microscopy (AFM), magnetic force microscopy (MFM), and lateral force microscopy (LFM). Most force microscopy techniques are variations of the same basic principle. Forces between the surface and a cantilever tip cause the tip to deflect up and down or sometimes side to side. Deflection of the cantilever shifts the position of a laser beam that reflects off the top of the cantilever onto a photodiode array. The movement of the beam between the photodiodes is used to calculate the cantilever deflection.

How is a scanning probe microscope useful in nanotechnology research and in nanofabrication?
The technique is primarily used to map surface forces at the nanoscale, but the SPM probes have sometimes been used to move objects such as atoms or "write" with molecular "ink."

How is the scanning probe microscope different from an atomic force microscope?
An atomic force microscope can be considered one type of scanning probe microscopy. Several other types of SPM exist (see: http://en.wikipedia.org/wiki/Scanning_probe_microscopy)

How did you and other researchers conceptualize the idea of building a model of a Scanning Probe Microscope from LEGO® building blocks? Did you try other building materials as well?
A number of people have used different materials to demonstrate SPM, including wooden sticks, hacksaw blades, and flexible sheet refrigerator magnets. We opted to use LEGO® bricks for a number of reasons.

First, many people are familiar with *LEGO®* bricks, and most models can be built with a level of mechanical sophistication that does not intimidate or frustrate the user. Second, *LEGO®* bricks typically have many connection points, allowing tremendous flexibility in the structures that can be built. A set of bricks can be used to model structures of matter and the techniques used to study them.

You have taught students how to build a scanning probe microscope model from *LEGO®* blocks. How old were the students who built the *LEGO®* model and how long did it take them to build it? Were there any problems in building the model?
Eighth graders as well as graduate students have built various creations of models. The simplest models can be built within a half hour. More complicated models take longer to build. The most challenging part of building the model appears to be mounting the laser pointer.

If a teacher or student wanted to construct the *LEGO®* model of the scanning probe microscope, whom should they contact for directions and for a list of materials?
Most of the directions for making the model are in the online book, "Exploring the Nanoworld with LEGO® Bricks," which can be accessed from http://mrsec.wisc.edu/Edetc/LEGO/index.html.

My contact information given on the site is:

Dr. Dean Campbell, Department of Chemistry, Bradley University, 1501 West Bradley University, Peoria, IL 61625. Phone: (309) 677-3029. E-mail: campbell@bradley.edu

What would your advice be to young people who would be interested in a career in nanotechnology? What are some of the opportunities in this field?
Nanotechnology is a hot field now and will likely remain so for some time. Nanotechnology is also a very interdisciplinary field. Successful

individuals will need to have a breadth of knowledge in mathematics, physical sciences and biology, as well as the ability to communicate effectively with others.

What are some of the benefits of nanotechnology and what would be some of the risks?
Many technologies stand to benefit from nanotechnology, ranging from computers to medicine to sunscreen. There are some concerns that nanoscale chemicals will have different properties and therefore different toxicities than bulk materials. Therefore, nanostructures must be carefully assessed for potential unexpected hazards.

NANO ACTIVITY: MODELING A SCANNING PROBE MICROSCOPE

In this activity, Professor Campbell provides background information about Scanning Probe Microscopy (SPM) and ideas on how to build a LEGO® model of it.

Background Information. *Scanning Probe Microscopy*

You are likely familiar with microscopes that use light, but many objects like atoms are too small to be seen using light. Other microscopes use electrons to view even smaller objects, but these also have difficulty visualizing objects as small as atoms.

Scanning probe microscopy (SPM) is a method for mapping surfaces of materials on the atomic scale. It is useful to study the locations of atoms on surfaces because many chemical reactions, such as corrosion and catalysis, take place at solid surfaces. Most SPM techniques are variations of the same basic principles, illustrated in the picture. At the heart of the microscope is a probe called a cantilever. The cantilever is fixed at one end and can flex up and down at the other end like a diving board. A laser beam is shined onto the tip of the cantilever. The light bounces off the top of the cantilever onto a pattern of light sensors called a photodiode array.

To run the SPM, the cantilever is brought very close to a surface. The surface is moved back and forth under the cantilever, so the cantilever more or less scans the surface. Forces between the surface and a cantilever tip cause the tip to deflect up and down. Deflection of the cantilever shifts the position of the laser beam that reflects off the top of the cantilever onto a photodiode array. The movement of the beam between the photodiodes is used to measure the amount of cantilever deflection. A computer combines the information about the cantilever deflection with information about the back and forth movement about the surface to produce a three-dimensional map of the surface. This SPM technique is so sensitive that individual atoms can deflect the cantilever probe and therefore be detected!

Directions for Building the Model

Most of the directions for making the model can be accessed at our Web site "Exploring the Nanoworld with LEGO® Bricks," http://mrsec.wisc.edu/Edetc/LEGO/index.html.

To get an idea of how your model will look when finished, observe the triangular LEGO® model in the photo. This model contains a laser pointer and a cantilever with a triangular LEGO® probe on its underside and a mirror atop it. In this model, the cantilever tip is in physical contact with a LEGO® surface. The LEGO® pegs on the surface deflect the cantilever. Light from the pocket laser reflects from the mirror on the cantilever and shines onto a wall.

After you build your model, have someone attach small bricks to the LEGO surface and then place the SPM behind a screen. Using the laser beam spot movement and horizontal surface movement, try to map the locations of the small bricks on the surface. Sketch your map of the surface on a sheet of paper.

Placing 1x1 bricks on a flat plate "substrate" makes raised bumps to push the cantilever (and the laser beam spot) up and down. A "handle" can be attached to this substrate and used to map to positions of the bumps. Place the SPM and substrate on a sheet of paper. Move the substrate back and forth (with the handle always pointing in the same direction). When a bump pushes the beam spot up, make a mark on the paper at the end of the handle. Note that the map produced will be rotated 180° from the actual bump positions.

Questions to Explore

1. Will a bump on a flat surface move the laser spot up or down?
2. How does the distance the SPM is positioned from the wall affect how much the laser beam spot shifts?

READING MATERIALS

Drexler, Eric K. *Nanosystems: Molecular Machinery, Manufacturing, and Computation.* New York: John Wiley & Sons, 1992.

Fritz, Sandy. *Understanding Nanotechnology: From the Editors of Scientific American.* New York: Warner Books, 2002.

Hall, J. Storrs. *Nanofuture: What's Next For Nanotechnology.* Amherst, NY: Prometheus Books, 2005.

Newton, David E. *Recent Advances and Issues in Molecular Nanotechnology.* Westport, CT: Greenwood Press, 2002.

Scientific American (authors). *Key Technologies for the 21st Century: Scientific American: A Special Issue.* New York: W.H. Freeman & Co, 1996.

VIDEOS

Try the Simulator. To see a simulation of a scanning tunnel microscope go to: http://nobelprize.org/educational_games/physics/microscopes/scanning/

Videos from the Hitachi Corporation, http://www.hitachi.com/about/corporate/movie/ What's Next in Nanotechnology?

Penn State University. *Amazing Creatures with Nanoscale Features*: This animation is an introduction to microscopy, scale, and applications of nanoscale properties. This activity is available for use via the Center Web site at http://www.cneu.psu.edu/edToolsActivities.html

Electron-Beam Lithography. Nanopolis Online Multimedia Library. Electron-beam lithography is a technique for creating extremely fine patterns required for modern electronic circuits. http://online.nanopolis.net/viewer.php?subject_id=139

NanoManipulator: Seeing and Touching Molecules. http://www.nanotech-now.com/multimedia.htm

Video Zyvex. Nanomanipulator, Zyvex S100 DVD Preview. http://www.zyvex.com/Research/SEM_manip/Manip.html

WEB SITES

Scanning Tunneling Microscope (STM): Describes the STM's development. http://physics.nist.gov/GenInt/STM/stm.html

Scanning Probe Methods Group, University of Hamburg: Academic research group using scanning probe methods (SPM), emphasis on investigating the relationship between nanostructure and nanophysical properties. http://www.nanoscience.de/

IBM Almaden STM Molecular Art: Some of the famous images of atoms and molecules made with IBM's scanning tunneling microscope. http://www.almaden.ibm.com/vis/stm/lobby.html

NanoManipulator: University of North Carolina—The NanoManipulator provides an improved, natural interface to SPMs (STMs and ATMs). http://www.cs.unc.edu/Research/nano/index.html

Exploring the Nanoworld. http://mrsec.wisc.edu/edetc.

The Incredible Shrunken Kids. http://www.sciencenewsforkids.org/articles/20040609/Feature1.asp

SOMETHING TO DO

You can create an edible, layered cookie (nanosmore) that will represent the process of photolithography by creating a patterned silicon wafer using a substrate and a photoresist with simple foods. www.nbtc.cornell.edu/mainstreetscience/nanosmores_and_photolithography.pdf

4

Carbon Nanotubes, Nanowires, and Nanocrystals

Carbon is one of the most abundant elements in living things and materials derived from living things. By volume, carbon is about the fourth most abundant element in the universe.

As you learned in Chapter 2, elements are types of atoms that make up all the things around us. Carbon plays an important role in nanotechnology research and for potential nanotechnology applications. Before we discuss the role of carbon as nanoparticles, let's review some general information about the chemical and physical properties of carbon in the macro-world.

THE ELEMENT CARBON

The chemical element carbon (C), is found in many different compounds. It is in the food we eat, the clothes we wear, the cosmetics we use, and the gasoline that fuels our cars. Carbon plays a dominant role in the chemistry of life and without carbon we could not exist.

Carbon is a naturally occurring nonmetallic element that is present in the cells of all organisms. Carbon has an atomic number of 6 and an atomic weight of 12. The carbon atom contains six neutrons and six protons and combines with other elements forming a variety of compounds that include carbon dioxide and carbon monoxide.

Carbon makes up about 19 percent of the mass of the human body and is an essential component of proteins, carbohydrates, fats, and nucleic acids. All fossil fuels, coal, petroleum, oil shale, and natural gas, contain carbon as a principal element and combined with hydrogen form a category of hydrocarbons.

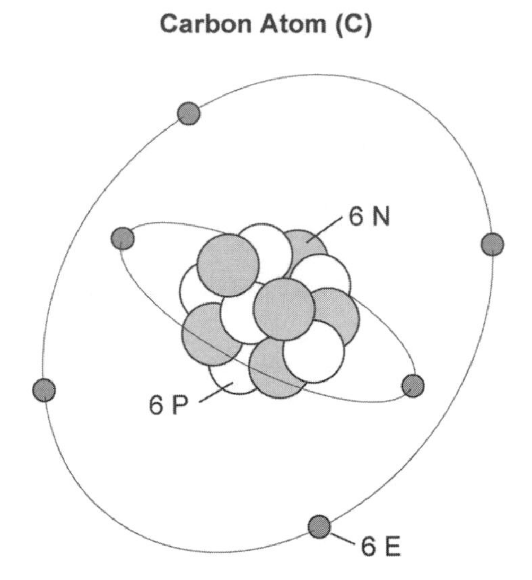

Figure 4.1 Carbon Atom. (*Courtesy of Jeff Dixon*)

Products Made from Carbon

As noted, carbon can combine with other elements as well as with itself. This ability allows carbon to form many different compounds of varying size, strengths, and shape. Carbon is used in synthetic materials called composites that contain both carbon fiber and fiberglass reinforcement.

Light yet strong, carbon composites have many applications in aerospace and automotive fields, as well as in sailboats, and notably in modern bicycles. Carbon is also becoming increasingly common in small consumer goods as well, such as laptop computers, tripods, fishing rods, tennis racket frames, stringed instrument bodies, classical guitar strings, and even drum shells.

Some Uses of Carbon

Graphite combines with clays to form the "lead" used in pencils.

A diamond is used for decorative purposes, and also as drill bits.

Carbon is added to iron to make steel.

Carbon in the form of graphite is used for control rods in nuclear reactors.

Graphite carbon in a powdered cake form is used as charcoal for cooking, artwork, and other uses.

Charcoal (carbon)is a constituent of gunpowder.

Charcoal pills are used in medicine in pill or powder form to adsorb toxins or poisons from the digestive system.

Different Forms of Carbon

There are several different allotropes or forms of carbon. Each form has a different molecular structure. The forms include amorphous, graphite, diamond, and fullerene.

The term amorphous means lacking a definite form or having no specific shape. In chemistry, the term amorphous means that the substance is lacking a crystalline form. For example, soot is an amorphous carbon that is a dark powdery deposit of unburned fuel residues. Glass, amber, wax, rubber, and plastics are other examples of amorphous substances. Amorphous carbon has been found in comets.

However, the other forms of carbon—diamonds, graphites, and fullerenes—have definite crystalline forms. Graphite is a very soft mineral consisting of loosely bond atoms arranged in a two-dimensional crystalline form that looks like a thin flat plane. This property makes graphite a good natural lubricant because it's flat, planar structure allows the nanoscale sheets of graphite to slide past one another easily. Although the mineral is soft, graphite's strength and its ability to conduct electricity makes the mineral especially useful in nanotechnology.

There are two main classifications of graphite, natural and synthetic. Graphite minerals can be found naturally in the Earth's crust. Synthetic graphite is made from petroleum coke.

Besides pencil production, graphite is also used to manufacture crucibles, ladles, and moulds for containing molten metals. Graphite is mainly used as electrical material in the manufacturing of carbon brushes in electric motors. Highly pure graphite is used in large amounts for the production of moderator rods and other components in nuclear reactors. The moderator in the nuclear reactor is used to slow down the neutrons so that the right speed is maintained for a steady fission rate.

Some other uses of graphite include:

- Aerospace applications
- Batteries
- Carbon brushes
- Graphite electrodes for electric arc furnaces for metallurgical processing.
- Graphite lubricants

Diamond

The hardest known natural mineral is diamond. The diamond contains carbon atoms that are stacked or arranged in a three-dimensional form or array. This structure makes the diamond super hard for the cutting and grinding of metals and other materials. Unlike graphite, a diamond is not a good conductor of electricity but the mineral is an excellent thermal conductor.

Applications of diamonds are used in a variety of products that include laser diodes and small microwave power device, integrated circuit substrates, and printed circuit boards. Diamond is starting to be used in optical components, particularly as a protective coating for infrared optics in harsh environments.

Diamond provides an impressive combination of chemical, physical, and mechanical properties: Some of these properties include:

- Low coefficient of friction
- High thermal conductivity
- High electrical resistance
- Low thermal expansion coefficient

FULLERENES AND NANOTECHNOLOGY

Fullerenes are the third allotropic form of carbon material (after graphite and diamond). Fullerenes are large molecules of carbon that are arranged in a form that looks much different from the shapes of graphite or diamond. Fullerenes are arranged in a form that is spherical, ellipsoid, or cylindrical. Fullerenes are about 1 nanometer in diameter. This compares to 0.16 nanometer for a water molecule.

> **Did you know?**
> Another allotrope of carbon is a spongy solid that is extremely lightweight and, unusually, attracted to magnets. The inventors of this new form of carbon—a magnetic carbon nanofoam—say it could someday find medical applications.

Fullerenes were discovered during laser spectroscopy experiments at Rice University in September 1985. The 1996 Nobel Prize in Chemistry was awarded to Professors Robert F. Curl, Jr., Richard E. Smalley, and Sir Harold W. Kroto for their discovery. Fullerenes were named after Richard Buckminster Fuller, an architect known for the design of geodesic domes, which resemble spherical fullerenes in appearance.

BUCKYBALLS

By far the most common and best-known fullerene is the buckminsterfullerene, buckyball, or C_{60}. It has a soccer-ball-shaped structure that includes 20 hexagons and 12 pentagons. Scientists have now discovered other buckyballs as well. They include C_{70}, C_{76}, and C_{84}.

> To watch and listen to Sir Harold Kroto explain why he named the carbon cluster that he discovered, buckminsterfullerene, go to: http://online.nanopolis.net/viewer.php?subject_id=268]

Applications of Buckyballs

The fullerene family of carbon molecules possesses a range of unique properties. A fullerene nanotube has tensile strength about 20 times that of high-strength steel alloys, and a density half that of aluminum. So it is stronger than steel and lighter than aluminum.

> **Did you know?**
> Rice University scientists had constructed the world's smallest car—a single molecule "nanocar" that contained a chassis, axles, and four buckyball wheels. The entire car measured just 3–4 nanometers across, making it slightly wider than a strand of DNA.

Since the discovery of fullerenes, scientists have discovered some possible uses for these molecules. Some of these potential applications include being used in making computers, fuel cells, and sensors.

Major pharmaceutical companies are exploring the use of fullerenes in drugs to control the neurological damage of such diseases as Alzheimer's disease and Lou Gehrig's disease (ALS). Companies are also testing the use of fullerenes in drugs for atherosclerosis and for use in antiviral agents.

A group of medical researchers believe that fullerenes could be used in tiny special sponges that would soak up dangerous chemicals from any tissues in the brain that have been injured. The sponges would immobilize the dangerous chemicals that would, if left untreated, destroy the nerve cells.

CARBON NANOTUBES

Scientists discovered that if you can make buckyballs big enough, they could become carbon cylinders called carbon nanotubes. Carbon nanotubes are long, thin cylinders of carbon molecules. A carbon nanotube is a completely different material from either diamond or graphite.

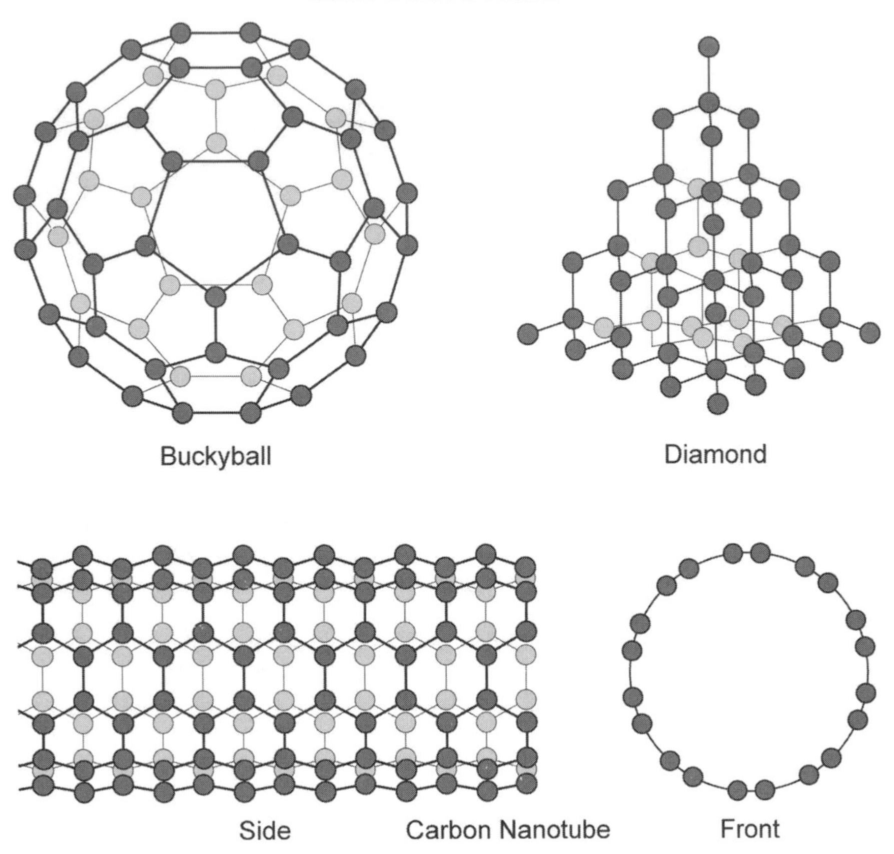

Figure 4.2 Carbon Forms. (*Courtesy of Jeff Dixon*)

Sumio Iijima, of NEC Corporation came upon carbon nanotubes in 1991 in Tsukuba, Japan, while researching buckyballs using an electron microscope. The carbon nanotubes he discovered can be visualized as a two-dimensional sheet of graphite. The arrangement of carbon atoms into a hexagonal lattice is called "graphene," because it has the form of a graphite sheet rolled into a cylinder. The nanotube looks like a rolled-up piece of chicken wire.

Carbon nanotubes have unique properties that make them potentially useful in a wide variety of applications such as in nanoelectronics, optics, and materials applications. They exhibit extraordinary strength and unique electrical properties, and are efficient conductors of heat.

Two Types of Nanotubes

There are two main types of nanotubes: single-walled nanotubes (SWNTs) and multiwalled nanotubes (MWNTs).

Most SWNTs have a diameter of close to 1 nanometer, with a tube length that can be many thousands of times longer. In fact, SWNTs can reach the length of centimeters. The structure of a SWNT can be conceptualized by wrapping a one-atom-thick layer of graphite called graphene into a cylinder. A graphene is a two-dimensional single sheet of carbon-bonded atoms.

A single-walled nanotube is commonly called a buckytube and consists of one single shell. The nanotube is cylindrical, with at least one end typically capped with a hemisphere of the buckyball structure. The diameter of a nanotube is only a few nanometers wide and can extend up to 50 microns in length.

Nanotubes have the following physical, chemical, and mechanical properties that make them such an outstanding material:

- Electrical conductivity: Depending on their precise structure, carbon nanotubes can be either metallic conductors or semiconductors. The electrons in nanotubes can travel much faster than in metals such as copper, and they do not dissipate or scatter. The electrical conductivity of the nanotubes could be useful in absorbing static noise, storing energy, or in replacing silicon circuits in computer chips.
- Thermal conductivity: The thermal conductivity of nanotubes is superior to that of a diamond. In some tests, nanotubes have been shown to have a thermal conductivity at least twice that of diamond. The nanotubes could be potentially handy for cooling off confined spaces inside computers and other nanoelectronics.
- Mechanical: Nanotubes are the stiffest, strongest, and toughest fiber known. For their small size, nanotubes are six times lighter than steel but more than 500 times stronger. They could be used to replace copper wires or to create superstrong plastics.

MWNTs are multiwalled nanotubes which may have 7–20 concentric grapheme cylinders. Double-walled carbon nanotubes have higher thermal and chemical stability than single-walled carbon nanotubes. MWNT can be applied to gas sensors, nanoelectronic devices, and nanocomposites.

HOW ARE CARBON NANOTUBES MADE?

There are several basic methods by which carbon nanotubes are made. Scientists can make modifications to each technique to suit their specific research purpose.

Laser Ablation

A high-power laser is used to vaporize carbon from a graphite target at high temperature. The resulting "soot" is collected by a water-cooled collector. This method is used to form single-walled nanotubes.

Electric Arc Discharge

In the electric arc discharge method, an electric current flows between two graphite rods. One graphite rod acts as an anode (a negative terminal from which electrons flow) and the other rod acts as a cathode (a positive terminal to which electron flow).

During the discharge, a hot, bright arc of electricity vaporizes carbon from the anode and produces plasma of carbon. The carbon condenses on the cathode to form nanotubes. This method produces mostly multi-walled nanotubes.

> **Did you know?**
> In physics, plasma is another state of matter. It is an ionized gas. This means the gas is heated or electrically charged to the point that it gives off light, as in a fluorescent light tube or bulb. Plasma screen televisions use this principle; they contain gas that emits light when charged. Essentially, the sun, like most stars, is a great big ball of plasma. In life science, a different substance called plasma is a part of blood.

Chemical Vapor Deposition (CVD)

In this method of making carbon nanotubes, metal catalyst particles are placed on a surface, such as a silicon wafer, and heated to high temperatures in a hydrocarbon gas. The high temperature and the catalyst particles break apart the hydrogen and the carbon atoms in the gas. A nanotube grows out from the catalyst and grows longer and longer as more carbon atoms are released from the gas. This method produces both multiwalled and single-walled nanotubes depending on the temperature.

APPLICATIONS OF CARBON NANOTUBES

Carbon nanotubes are the driving force for current advances in nanotechnology; they have excellent mechanical and electronic properties,

Table 4.1 Some Potential Applications for Nanotubes

Actuators/Artificial Muscles
Atomic Force Microscope Probe Tips
Batteries
Chemical Sensors
Collision-Protection Materials
Controlled Drug Delivery
Data Storage
Dialysis Filters
Flat Panel Display Screens
Flywheels
Hydrogen Storage
Microelectromechanical (MEMS) Devices
Microelectronics/Semiconductors
Nanoelectronics
Nanogears
Nanolithography
Nanothermometers
Nanotweezers
Reinforcement of Armor
Solar Storage
Super capacitors
Thermal Protection
Waste Recycling

the latter making them extremely attractive for electronics. Being only a few times wider than atoms, the carbon nanotubes offer exceptionally high material properties, such as electrical and thermal conductivity, stiffness, toughness, and remarkable strength.

Sheets of carbon nanotubes have the following features:

- The sheets are transparent and stronger than steel or high-strength plastics and can be heated to emit light.
- A square mile of the thinnest sheets, about 2-millionths-of-an-inch thick, would weigh only about 170 pounds.
- In lab tests, the sheets demonstrated solar cell capabilities for using sunlight to produce electricity.

Other potential applications of carbon nanotubes include the following.

Flat Panel Display Screens

When a nanotube is put into an electric field, it will emit electrons from the end of the nanotube like water being pushed through a high-powered water hose. If the electrons bombard a phosphor screen, then an image can be created. Several companies are researching how to use this nanotube technology to replace the bulky electron guns of conventional TV sets with smaller carbon nanotube electron guns.

Nanoscale Electronics and Carbon Nanotubes

Many nanoelectronic companies are interested in developing new methods to manufacture transistors. Transistors are the key component of the electrical circuit that is used in the operation of computers, cellular phones, and other electronic devices. In fact transistors are used in nearly every piece of electronic equipment today.

The transistors are a key building block of electronic systems—they act as bridges inside the computer chip that carry data from one place to another. The more transistors on a chip, the faster the processing speed.

To manufacture more transistors on a chip, several companies are now experimenting on how to make the channel length in the transistors smaller and smaller. The channel in the transistor is the path where data travels from one place to another inside chips. Some success in this area has already happened. One company has successfully used carbon nanotubes to make smaller channels in the production of their transistors.

> To see a carbon nanotube form a channel, go to: http://online.nanopolis.net/viewer.php?subject_id=268]

The achievement of making smaller channels in the production of transistors is an important step in finding materials, such as carbon nanotubes, to be used to manufacture computer chips. Moore's Law states that the number of transistors that can be packed on a chip doubles every 18 months. But many scientists expect that within 10–20 years the present computer chip made with silicon will reach its physical limits. The ability to pack more transistors on the chip will no longer be feasible. So the application of carbon nanotubes in transistors may help increase the storage problem on a chip.

> **Did you know?**
> Moore's Law was a statement by George E. Moore, cofounder of Intel. Moore stated that the complexity of these circuits would double every 2 years.

Sciencenter of Ithaca, New York has a traveling nanotech exhibit called *It's a Nano World* and another called *Too Small To See.* Attending the exhibits, school-aged children can learn about tiny things by walking through and playing with very large and colorful things in a traveling science museum exhibition. In this photo, a group of students are building a carbon nanotube. (*Courtesy Gary Hodges and the Sciencenter, Ithaca, NY*)

AFM PROBE TIPS

Single-walled carbon nanotubes have been attached to the tip of an AFM probe to make the tip "sharper." This allows much higher resolution imaging of the surface under investigation; a single atom has been imaged on a surface using nanotube-enhanced AFM probes. Also, the flexibility of the nanotube prevents damage to the sample surface and the probe tip if by accident the probe tip slams into the surface.

Hydrogen Fuel Cells and Nanotubes

There is a global effort today to develop renewable energy sources. A great deal of research is being devoted to the production of hydrogen fuel cells. A fuel cell is an electrochemical energy-conversion device. Most fuel cells in use today use hydrogen and oxygen as the chemicals. When oxygen and hydrogen react in a fuel cell electricity is produced and water is formed as a by-product.

To use a hydrogen-oxygen fuel cell to generate electricity, scientists and engineers must find a safe way to store hydrogen gas in the fuel tank. Hydrogen gas takes up a lot of volume. Carbon nanotube materials built into the inside of the tank may be an excellent option.

Carbon nanotubes are extremely porous. So, acting like a sponge, the carbon nanotubes in the tank would absorb large capacities of hydrogen gas being pumped into the tank. By absorbing large quantities of hydrogen gas, the carbon nanotubes would allow more hydrogen gas to be stored in a tank. In this way, you get a fuller tank of hydrogen gas than if you pumped the same amount of hydrogen gas into a noncarbon nanotube tank.

Did you know?
Hydrogen (Atomic Number 1) is the lightest element. It is by far the most abundant element in the universe and makes up about 90 percent of the universe by weight. Hydrogen as water (H_2O) is absolutely essential to life and it is present in all organic compounds.

While nanotube-based hydrogen cells for automobiles look promising, it may be awhile before any large numbers of hydrogen fuel cell automobiles are on the road.

Actuators/Artificial Muscles

An actuator is a device that can provide motion. In a carbon nanotube actuator, electrical energy is converted to mechanical energy causing the nanotubes to move.

Here is how it works. Two pieces of paper made from carbon nanotubes are placed on either sides of a piece of tape that is attached to an electrode. An electrode is a conductor through which electric current is passed.

When electric current is applied the electrons are pumped into one piece of the paper. The nanotubes on that side of the paper expand causing the tape to curl in one direction. This curling movement is similar to the way an artificial muscle works. The nanotube actuator can produce 50 to 100 times the force of a human muscle the same size. Researchers see the possible applications of the nanotube actuator for robotics and prosthetics.

Nanotechnology in Chemical Sensors

Sensors are used in our everyday life even though they seem to be out of our sight. There are motion sensors, acoustic sensors, electrical power sensors, distance sensors, mechanical sensors, and chemical sensors. Sensors are used in automobiles, machines, aircraft, medicine, industry, and robotics.

Nanotechnology researchers are now developing and improving upon chemical sensors, using nanoparticles. Chemical sensors detect the presence of very small amounts of specific chemical vapors or classes of chemicals.

You are probably familiar with a carbon monoxide detector. A carbon monoxide detector is a chemical sensor often used in the home. These detectors continually sample the air in the rooms and will sound an alarm if the invisible and odorless carbon monoxide levels become dangerous in the home and workplace.

Researchers now want to develop small, inexpensive sensors that can sniff out chemicals just as dogs do when they are used in airports to smell the vapors given off by explosives or drugs hidden in packages or in other containers. These small and inexpensive sensors can be placed throughout an airport, shopping mall, or building where security measures are needed, to check for vapors given off by explosive devices. These sensors can also be useful in industrial plants that use chemicals in manufacturing to detect the release of chemical vapors.

Using a chemical sensor to detect a chemical vapor is fine, but you want to be able to identify and analyze the gas vapor quickly. You want to know what it is. Is it OK to breathe? Is it dangerous? In 2005, researchers at the Naval Research Laboratory were able to detect minute amounts of sarin gas in under 4 seconds using a prototype nanotube gas sensor. Previous sensors

Did you know?

Chemical sensors are used to improve quality control in the processing of various molten metals, including steel, aluminum, and zinc.

took over a minute to detect the same amount. If you are unfamiliar with sarin, it is known by its NATO designation of GB (O-Isopropyl methylphosphonofluoridate), an extremely toxic substance whose only application is as a nerve agent. As a chemical weapon, it is classified as a weapon of mass destruction by the United Nations. In 1995, there was a sarin gas attack on a Tokyo, Japan, subway that killed several people and injured many others.

In the future, nanotube chemical sensors can play an important role in security and environmental applications.

Carbon Nanotubes and NASA

Carbon nanotubes may offer the possibility of detecting harmful ionizing radiation during interplanetary space missions. Radiation damages DNA in living cells, leading to health problems such as nausea, cataracts, and cancer.

Radiation hazards in space travel come in the form of trapped radiation, galactic cosmic rays, and solar particle events. To detect and monitor the radiation, NASA is developing a carbon nanotube dosimeter. A dosimeter is any device used to measure an individual's dose to a hazardous environment. The carbon nanotube dosimeter would detect and measure the radiation by monitoring changes in the conductivity of a nanotube sensor. Studies have shown that nanotube conductivity levels in the dosimeter increase with radiation and then decrease after that. An increase in radiation doses in a spacecraft would warn the astronauts to take action.

NASA Space Elevator

NASA is also planning a space elevator that would be attached to a cable that would orbit Earth at a height of 36,000 kilometers. Scientists believe that the cable could be made from carbon nanotubes that are 100 times stronger than steel. See Chapter 8 for more information about NASA's plans for the space elevator.

NOT ALL NANOTUBES ARE CARBON

One company has found a way to make commercial quantities of naturally occurring nanotubes that are not carbon. The halloysite nanotubes, containing the elements aluminum, silicon, hydrogen, and oxygen, are found in halloysite clay found in the soil in certain areas in the country. Halloysite clay has historically been used to make fine china and ceramics.

The company, NaturalNano, is developing methods of processing the halloysite clay into nanotubes for use in commercial applications. The halloysite nanotubes can be used in such products as additives in polymers and plastics, electronic components, cosmetics, and absorbents.

NANOWIRES, NANOCRYSTALS, AND QUANTUM DOTS

Nanowires

A nanometer-scale wire is made of metal atoms, silicon, or other materials that conduct electricity. Nanowires are built atom by atom on a solid surface.

A nanowire is a very small wire that is composed of either metals or semiconductors. It is also known as a nanorod or quantum wire since the dimensions of the nanowire are in the order of a nanometer (10^{-9} meters). The nanowires have the potential to be used as components to create electrical circuits.

Production of Nanowires

The fabrication of a nanowire can be done either with vapor deposition using specialty gases, or with a laser aimed at a target material to produce a specific vapor. The gases eventually condense on a substrate material, like silicon, forming a nanowire.

The tiny nanowires have the potential to be used in medical applications. A tiny nanowire sensor used in medical diagnostic tests is 1,000 times more sensitive than conventional tests. The nanosensor is capable of producing test results in minutes rather than in days or weeks. This feature could pave the way for faster and more accurate medical diagnostic tests that would allow earlier disease detection and intervention.

Several companies are developing nanowire devices. These devices will be used for chemical sensing, lasers and light-emitting diodes, and, in the future, nanoelectronics.

Nanocrystals

Nanocrystals are grown from inorganic materials, including metals and semiconductors. Some researchers have made nanocrystals of silver, gold, platinum, palladium, ruthenium, rhodium, and iridium. Nanocrystals are approximately 10 nanometers in diameter.

Some of the potential applications will include using nanocrystals as building blocks for producing strong metals and composites. The technology is also applicable to lighting, high-resolution imaging, and semiconductor materials. Since nanocrystals emit colored light, they will have a big impact on how everything from large-screen televisions to portable electronics are manufactured.

In fact, nanocrystals could be the next generation of photovoltaics. The major problems of solar cells are that they have low specific power of efficiency and they cost a lot. The nano-engineered solar cells have the potential to solve these issues. Using nanocrystals may be able to produce very low cost solar cells that generate energy at an installed photovoltaic system cost (capital cost of system) of less than under a dollar per watt. This would mean having a 5-KW system for less than $5,000, which would be competitive with fossil fuel prices.

> **Did you know?**
> Nanocrystals made with zeolite are used as a filter to turn crude oil into diesel fuel at a major oil refinery in Louisiana, a method cheaper than using the conventional way.

Quantum Dots

A quantum dot is a semiconductor nanocrystal that is about 1 to 6 nanometers in diameter. It has a spherical or cubic-like shape consisting of thousands of atoms,

A quantum dot is made of cadmium selenide (CdSe), cadmium sulfide (CdS), or cadmium telluride (CdTe) and then coated with a polymer. The coating is used to prevent these toxic chemicals from leaking.

The CdS is used for UV-blue, the CdSe for the bulk of the visible spectrum, and the CdTe for the far red and near infrared. The particle's size determines the exact color of a given quantum dot.

A wide range of colors can be emitted from a single material simply by changing the dot's size and makeup. A larger dot emits the red end of the spectrum and the smaller ones emit blue or ultraviolet. As an example, a cadmium selenide (CdSe) quantum dot more than 6 nanometers in diameter emits red light, while one less than 3 nanometers wide glows green.

> **Did you know?**
> The human eye can see radiation as colors ranging from red wavelengths of about 700 nanometers to violet wavelengths of about 400 nanometers.

Quantum dots could help scientists image the behavior of cells and organs to a level of detail never before seen. Conventional fluorescent dyes used in the life sciences to help researchers monitor how cells and organs grow and develop normally lose their ability to emit light within seconds. On the other side, quantum dots emit light far longer, helping scientists monitor cells and organs in diseased and healthy conditions.

The wide range of colors that can be produced by quantum dots makes them well suited for a variety of applications. As an example, they have great potential in security. Quantum dots could, for example, be embedded in banknotes or credit cards, producing a unique visible image when exposed to ultraviolet light. The image would identify the rightful owner of the credit card or banknote. Quantum dots could also be used in electronics applications such as data storage, light-emitting diodes, photovoltaic devices, flat-panel displays, and in medical aplications.

Quantum Dots and Cancer

Emory University scientists have used luminescent quantum dot nanoparticles in living animals to simultaneously target and image cancerous tumors. The quantum dots were first coated with a protective shell covering. Then special antibodies were attached to the surface of the quantum dots. After the quantum dots were injected into the body, they were guided to the prostate tumor of the living mice. Using a mercury lamp, the scientists were able to see the surface of the tumor illuminated by the accumulation of quantum dots on the cell. The scientists believe the ability to both target and image cells in vivo (in the body) represents a significant step in the quest to eventually use nanotechnology to target, image, and treat cancer, cardiovascular plaques, and neurodegenerative disease in humans. See Chapter 5 for more information about nanotechnology and nanomedicine.

Quantum Dots for Solar Cells

Scientists have been doing a lot of research with quantum dots to make photovoltaic cells more efficient. Presently, much of the solar energy striking photovoltaic cells is wasted. Here is why. When solar photons, fundamental light particles, strike a solar cell, they release electrons in the semiconductor to produce an electric current. However, when an electron is set free by the photon, it collides often with a nearby atom. The collision makes it less likely to set another electron free. So even though solar photons carry enough energy to release several electrons, producing more electricity, they are limited to one electron per solar photon, according to solar experts. As a result, most conventional solar cells operate at 15 to 20 percent efficiency using solar energy.

Researchers at the National Renewable Energy Laboratory (NREL) and the Los Alamos National Laboratory have been experimenting with quantum dots as a semiconductor in a solar cell. They have discovered that the use of the quantum dots allows solar energy to release multiple electrons, not just one. This research has the potential to make major improvements in the manufacturing of photovoltaic cells. The two research teams have calculated that a maximum efficiency of solar conversion at 42 percent efficiency is possible from the conversion of solar energy to electricity. So, more efficient solar cells are in the development stage.

NANOSHELLS

Nanoshells are a new type of nanoparticle composed of a substance such as a silica core that is coated with an ultrathin metallic such as a gold layer. Nanoshells are about 1/20th the size of a red blood cell and are about the size of a virus or about 100 nanometers wide. They are ball-shaped and consist of a core of nonconducting glass that is covered by a metallic shell, typically either gold or silver.

Nanoshells are currently being investigated as a treatment for cancer similar to chemotherapy but without the toxic side effects. These nanoshells can be injected safely into the body as demonstrated in animal tests. Once in the body, the nanoshells are illuminated with a laser beam that gives off intense heat that destroys the tumor cells.

In preliminary testing, one research medical team is using nanoshells combined with lasers to kill oral cancer cells. Oral cancer is a cancerous tissue growth located in the mouth. Smoking and other tobacco use are associated with 70 percent to 80 percent of oral cancer cases. Approximately 30,000 Americans will be diagnosed with oral or pharyngeal cancer each year. Human clinical trials using applications of nanoshells for cancer treatment will begin within a few years. However, nanoshells are already being developed for other applications. They include drug delivery and testing for proteins associated with Alzheimer's disease.

NANO INTERVIEW: PROFESSOR TIMOTHY SANDS, PH.D., PURDUE UNIVERSITY

Professor Timothy D. Sands is the Director of the Birck Nanotechnology Center in Purdue University's Discovery Park, located in West Lafayette, Indiana. He is also the Basil S. Turner Professor of Materials Engineering and Electrical & Computer Engineering at Purdue. Dr. Sands received his B.S., M.S., and Ph.D. degrees at UC Berkeley. Professor Sands and his research group investigate nanowires, nanotubes, and superlattice materials for energy conversion devices and nanoelectronics.

Where did you grow up and what were some of your favorite activities and subjects as a teenager?
I grew up and went to schools in northern California near San Francisco. As a young person, I enjoyed the outdoors and nature. I was interested in collecting insects, bird watching, and I was very active in the Audubon Society, as well. I hiked and explored the wetlands and hills in the San Francisco Bay Area.

What colleges did you attend and what was your major?
After I graduated from high school, I attended the University of California, Berkeley where I received all of my degrees including my Ph.D. degree in Materials Science.

What subjects do you teach at Purdue University?
I teach materials engineering, electrical engineering, and nanotechnology to undergraduates and graduate students.

What is materials engineering?
Materials engineering is the study of the relationships between composition, processing, microstructure, and properties of materials with the aim of improving the performance of materials that are used for everything from bridges to biochips.

What are your duties as the Director, Birck Nanotechnology Center?
My principal duty as the director is to facilitate the research projects at the Center. We have 45 resident faculty members and 200 resident graduate students. Since all of them come from a broad range of disciplines, as many as 30 disciplines, they need to learn how to work together. As an example, in our research projects, a biologist would be working with an engineer. So, my duties at the center are to facilitate the interdisciplinary work of the group projects.

What is your science background and how did you get interested in the field of carbon nanotubes and nanotechnology?
While at Berkeley as a student, I was able to take 50 percent of my classes in engineering and the other 50 percent in science courses. The science courses were mostly in the field of physics.

I was always interested in small things such as atoms in crystals and molecular structures. I did not get interested in carbon nanotubes until I ran into a colleague at Purdue. His name is Timothy Fisher, and he is a mechanical engineer and professor at Purdue University.

Tim is an expert in carbon nanotubes and one day we started talking about our interests. I was working on nanoporous materials and he was working on nanotubes. So, we decided to combine our interests and lab experiences to see if we could grow the carbon nanotubes inside a porous material. This experience got me interested in carbon nanotubes. While working as a team, we both also learned more about each other's fields of study.

When we use the term, *nanoelectronics*, what are we talking about?
Nanoelectronics to me really talks about bridging bottom-up nanofabrication with top-down nanofabrication to manufacture electronic systems. I think we are at the point where top-down nanofabrication is slowing down (running out of gas) and it is getting very expensive to use this method for manufacturing nanoelectronic circuits. So, now we need to learn ways to use the bottom-up nanofabrication approach in concert with top-down nanofabrication to make things at the nanoscale. This, of course, is a big challenge. The big opportunity is that carbon nanotubes have properties that are extraordinary, and if you can harness these properties, you would indeed be able to make higher performing nano devices and a higher performance chip.

How important are carbon nanotubes in the field of nanoelectronics?
Carbon nanotubes stand out because they may offer the biggest potential for nanotechnology breakthroughs. One reason is that the carbon nanotubes have high thermal conductivity, the highest of any known material. A major problem or bottleneck today in electronics is the removal of excessive heat in the active region of the chip. Single-walled carbon nanotubes could solve this problem because they would not overheat. Electrons also move readily in carbon nanotubes, which makes devices switch faster.

How did you and Professor Fisher develop a technique to grow individual carbon nanotubes vertically on top of a silicon wafer?
Growing nanotubes vertically allows us to stack more transistor circuits and other components in a computer chip, while keeping the same footprint as a conventional chip. This would be the electronic equivalent of a skyscraper.

We grow individual nanotubes vertically out of tiny cavities on top of a silicon wafer. This "vertically oriented" technique of growing carbon nanotubes on top of a silicon wafer might be an important step towards new kinds of computer chips with nanoelectronic devices, including wireless equipment and sensors.

Growing carbon nanotubes vertically on a silicon wafer may be a new way of constructing future microchips that are much faster to make and more energy efficient than conventional chips. Stacking the components on top of each other also cuts the distance and the time an electrical

Students Vijay Rawat and David Ewoldt and Professor Sands are using a special laser machine to deposit different types of multilayered nitride films with thickness precision at the nanoscale. The team plans to use the nitride films for thermoelectric applications. (*Photo credit: Birck Nanotechnology Center, Discovery Park, Purdue University*)

signal needs to travel in a microchip. But more research work needs to be conducted using this technique.

How do you grow the nanotubes?
We start with a porous material that is easy to make. The porous material is formed directly on the surface of a silicon wafer. In the process of making the porous material, we embedded a catalyst inside the pores. The catalyst particles are used to nucleate the nanotubes in a specific location. The wafer is placed in a plasma-enhanced deposition system that creates a carbon vapor that, when in contact with the catalyst particles, starts the growth of a carbon nanotube. The nanotubes self assemble and only one carbon nanotube forms per pore. There is no way a second nanotube will form in the same pore. The nanotubes grow vertically and eventually project out of the pores.

What tools and equipment did you use in growing the nanotubes? Did you need to work in a cleanroom environment?
We use all the state-of-the-art tools and equipment we can get to do our work. Students building the vertical carbon nanotubes use scanning

probe microscopes to measure the properties of the tubes. They use scanning electron microscopes to observe what they are doing because you cannot see these nanomaterials with a light optical microscope. Other techniques include using advanced lithography, a form of printing, to make patterns in a cleanroom environment. Students use an electron beam patterning system to write very fine lines, as narrow as about 10 nanometers, or about 50 atoms across. Lasers are used to measure properties of individual nanotubes to determine their semiconducting or metallic properties.

Does Purdue University collaborate with major computer companies?
There is much informal collaboration that goes on between universities and major computer chip corporations.

What advice would you give young people who would be interested in a career in nanotechnology?
Certainly, anyone thinking about nanotechnology should take the basic classes in biology, chemistry, physics, and mathematics. If students and teachers are interested in learning more about nanotechnology, I would recommend that they visit, **nanoHUB.org**., *the Web site of the Network for Computational Nanotechnology (NCN).* The NCN, led by Purdue, is a large network of many colleges and universities throughout the country and is funded by the National Science Foundation.

The original idea of the network was to put up simulation tools for researchers, but it turns out that the most popular section of the Web site is the tutorial section, attracting all levels of viewers, from high school students to those taking college courses.

The categories on the nanoHUB include *Simulate, Research, Teach and Learn, and Contribute.* The *Teach and Learn* site is a good place to start for someone who wants an introduction to nanotechnology. The topics include: *Nano 101, Nanocurriculum, Learning Modules,* and *Teaching Materials for K-12 and Undergraduates.*

NANO ACTIVITY: BUILDING BUCKYBALLS, A *NASA EXPLORES ACTIVITY*

Fullerenes, such as buckyballs, are being researched for several uses, including propellants, superconductors, lubricants, and optical equipment. In fact, *Science* magazine even elected the buckyball "molecule of the year" in 1991.

You can build a buckyball model by going to the *NASAexplores* Web site.

Here you will find a pattern and instructions on a PDF file for students and teachers.

The PDF file: http://www.nasaexplores.com/search.php

Procedure: Photocopy and then print out the buckyball figure

1. Cut around the perimeter of the buckyball template.
2. Cut along each dotted line to the star.
3. Cut out the shaded areas.
4. Beginning with the space labeled G in the upper right, place a drop of glue on the printed side. Slide this ring under the adjacent ring, making a five-sided hole surrounded by six-sided shapes.
5. Repeat step 4 until the sphere is formed.

Good luck.

READING MATERIAL

Hall, J. Storrs. *Nanofuture: What's Next For Nanotechnology.* Amherst, NY: Prometheus Books, 2005.

Krummenacker, Markus, and James J. Lewis. *Prospects in Nanotechnology: Toward Molecular Manufacturing.* New York: John Wiley & Sons, 1995.

Newton, David E. *Recent Advances and Issues in Molecular Nanotechnology.* Westport, CT: Greenwood Press, 2002.

Poole, Charles P., and Frank J. Owens. *Introduction to Nanotechnology.* Hoboken, NJ: John Wiley & Sons, Wiley-Interscience. 2003.

Regis, Edward. *Nano! The True Story of Nanotechnology—the Astonishing New Science That Will Transform the World.* United Kingdom, LONDON: Transworld Publishers Ltd., 1997.

VIDEOS

Nanoscale. Professor Wendy Crone What is a Nanoscale? Discusses Quantum Effects and Quantum Dots and Surface to Volume Ratio. Conversations in Science. Madison Metropolitan School District. UW-Madison Interdisciplinary Education Group. http://mrsec.wisc.edu/Edetc/cineplex/MMSD/index.html

Carbon Nanotube Transistors. Nanopolis Online Multimedia Library. The carbon nanotubes are ideal building blocks for molecular electronics, such as transistors. http://online.nanopolis.net/viewer.php?subject_id=268

Nanowires and Nanocrystals for Nanotechnology. Yi Cui is an assistant professor in the Materials Science and Engineering Department at Stanford. www.google.com/videoplay?docid=6571968052542741458

What is a buckminsterfullerene. Sir Harold Kroto. The Nobel Laureate explains why he named the carbon cluster that he discovered as a buckminsterfullerene.

http://www.invention.smithsonian.org/video/ and http://invention.smithsonian.org/centerpieces/ilives/kroto/kroto.html

Forming Carbon Nanotubes. University of Cambridge.

Two videos show how nickel reacts in a process called catalytic chemical vapor deposition. This film demonstrates one of several methods of producing nanotubes. Text accompanies the video for better understanding of the process. http://www.admin.cam.ac.uk/news/special/20070301/?

WEB SITES

National Science Foundation, Nanoscale Science and Engineering: http://www.nsf.gov/home/crssprgm/nano/start.htm

Nanodot: News and Discussion of Coming Technologies. http://nanodot.org/

The Nanotube Site: Michigan State University's Library of Links for the Nanotube Research Community. http://www.pa.msu.edu/cmp/csc/NTSite/nanopage.html

Evident Technologies: Manufacturer of quantum dot nanoparticles for use as a color-enhancing additive in optical devices and visual materials. http://www.evidenttech.com

NanoVantage Inc.: Portal offering monthly reports on new developments in nanotechnology. Emphasis on nanoparticles and nanocrystal materials. http://www.nanovantage.com/

SOMETHING TO DO

Make a paper model of a nanotube. www.stanford.edu/group/cpima/education/nanotube_lesson.pdf

5

Nanotechnology in Medicine and Health

One of the most promising applications of nanotechnology is nanomedicine. Nanomedicine is defined as the application and development of nanoscale tools and machines designed to monitor health care, deliver drugs, cure diseases, and repair damaged tissues. Nanomedical research will be an essential tool to diagnose, treat, and to do follow-up care in major diseases such as cardiovascular diseases, cancer, diabetes, and other diseases.

CARDIOVASCULAR DISEASES

Cardiovascular diseases are the most frequent cause of death in the United States, Europe, and the world, according to the World Health Organization. In the United States, cardiovascular disease accounts for twice as many deaths as all cancers in the country. Over 13 million people in the United States have coronary heart disease (CHD). Americans suffer approximately 1.5 million heart attacks annually and about half of them prove fatal, according to medical researchers.

To help diagnose and treat heart patients, one group, The National Heart, Lung, and Blood Institute (NHLBI) and the National Institutes of Health (NIH) have awarded researchers from Georgia Institute of Technology and Emory University $11.5 million to establish a new research program focused on creating advanced nanotechnologies to analyze plaque formation on the molecular level and to detect plaque at its early stages. Plaques, containing cholesterol and lipids, may build up during the life of blood vessels. When these plaques become unstable and rupture they can block the vessels, leading to heart attack and stroke.

The NHLBI researcher's programs will focus mostly on detecting plaque and pinpointing its genetic causes. The scientists will use three types of nanostructured probes. They include molecular beacons, semiconductor quantum dots, and magnetic nanoparticles.

> **Did you know?**
> Biologists categorize the molecules of living organisms into lipids, carbohydrates, proteins, and nucleic acids.

A molecular beacon is a biosensor. A biosensor is an analytical instrument capable of detecting biological molecules. The molecular beacon, only 4 to 5 nanometers in size, will seek out and detect specific target genes in the cells. Each cell in the human body contains about 25,000 to 35,000 genes that determine your traits. Scientists are studying genes to determine what illnesses genes cause.

The light emitted from the beacon will create a glowing marker if the cell has a detectable level of a gene that is known to contribute to cardiovascular disease.

The second type of probe is the semiconductor quantum dots. This is also used to study the molecules of cardiovascular disease. Quantum dot-based probes can be used to study interactions in live cells or to detect diseased cells. These ultrasensitive probes may help cardiologists understand the formation of early stage plaques and dramatically improve detection sensitivity.

The last probe in their research will include the nanostructured probe—magnetic nanoparticles. This probe will detect early-stage plaques in patients. The magnetic nanoparticles will target the surface of cells in a plaque and provide an image of the plaque formation. This technique could become a powerful tool for better disease diagnosis.

What Causes Cardiovascular Diseases?

One of the major causes of cardiovascular disease in most cases is the formation of plaque in the blood vessels. The plaque can lead to the clogging of the blood vessels causing death, disability, or strokes. Strokes are caused by ruptured blood vessels leaking blood into the brain.

Here are some symptoms of stroke:

- Sudden numbness or weakness of the face, arm or leg, particularly if it is on one side of the body
- Sudden confusion, trouble speaking or understanding. Sudden difficulty with walking, dizziness, loss of balance or coordination

- Sudden trouble seeing from one or both eyes
- Sudden severe headache with no known cause.

Nanoparticles Break Down Blood Clots

Strokes and other kinds of heart attacks are also caused by the formation of blood clots in the circulatory system. Blood clots can lead to a range of serious medical conditions, including heart attacks, pulmonary embolisms, and strokes. The main component of a clot is the insoluble fibrin. Fibrin is a blood-clotting protein and has a role in normal and abnormal clot formation (coagulation) in the body.

One research group is experimenting with different ways of treating fibrin. One test involves the use of nanoparticle drugs that dissolve and break up the fibrin. Preventing blood clots reduces the risk of stroke, heart attack, and pulmonary embolism.

Another research team is experimenting with a cardiovascular method that employs specially made molecules called "nanolipoblockers." The nanolipoblockers prevent cholesterol from causing inflammation and plaque buildup at specific blood vessel areas in the body.

> New Technology Helps Diagnose Heart Disease. Grouper. Go to: http://grouper.com/video/MediaDetails.aspx?id=1869316

In preliminary tests, these molecules show good results against harmful cholesterol.

Heart Stents and Nanotechnology

Doctors have used special stents for heart patients to open arteries that have become too narrow due to atherosclerosis. Atherosclerosis is the cholesterol plaque that builds up on the inner walls of arteries. As noted earlier, the plague prevents the blood vessels from carrying oxygen-rich blood throughout the body. The blockage of the blood flow can lead to a stroke or heart attack.

During or after heart surgery, a stent is inserted permanently into an artery. The stent, the size of a small metal tube, keeps the artery open, which improves blood flow, and helps prevent buildup of plague. The stent also enables the arteries to grow new tissue after the vessel-clogging plaque deposits have been removed.

However, some problems can occur with stents. The stent placement in the body can cause infection, blood clots, or bleeding. One major problem is that the body's immune system can reject the metal stents. When this happens, the immune system can cause the creation of scar tissue. In some cases, the scar tissue can build up inside blood vessels and interfere with blood flow.

Now, nanoresearchers are using a variety of methods to find new materials that cause the cells to attach better to these stents without creating as much dangerous scar tissue. One research company is testing a nano-thin coating application that is designed to protect surrounding tissue from any potentially harmful interactions with metal stents.

Stroke Stopper. Neuroradiologists Treat Brain Strokes with New Kind of Stent, Science Daily Video. Go to: http://www.sciencedaily.com/videos/2006-04-07/

CANCER DETECTION AND DIAGNOSIS

Cancer is currently the second leading cause of death in the United States and Europe. Detection of cancer at an early stage is a critical step in improving cancer treatment. Currently, detection and diagnosis of cancer usually depend on changes in cells and tissues that are detected by a doctor's physical exam or by laboratory-generated images on film. Scientists would like to make it possible to detect cancer at the earliest changes in cells or tissues before a physical exam.

Cancer

To detect cancer at its earliest stages, the National Cancer Institute (NCI) has invested 144 million dollars to develop and apply nanotechnology to cancer. The NCI envisions that within the next 5 years nanotechnology will result in significant advances in early detection, molecular imaging, assessment of therapeutic methods, and prevention and control of cancer. Cancer is an abnormal growth of cells anywhere in the body. It occurs when the genes in a cell allow it to split (make new cells) without control. There are many kinds of cancer, because there are many kinds of cells in the body. Some cancers form growths called tumors, while others, like cancers of the blood (leukemias), travel all over the body. Cancers may harm the body in two ways. They may replace normal cells with cells that do not work properly and they may kill normal cells.

Nanotechnology offers a wealth of tools that are providing cancer researchers with new and innovative ways to diagnose and treat cancer. Already, nanotechnology has been used to create new and improved ways to find small tumors through imaging. Nanoscale drug delivery devices are being developed to deliver anticancer therapeutics specifically to tumors.

Nanotechnology provides opportunities to prevent cancer progression. For example, nanoscale systems, because of their small dimensions, could be applied to stop progression of certain types of breast cancers.

Examples of nanotechnology in cancer research today include the following:

- Nanoscale cantilevers and nanowire sensors that can detect a cancer from a single cell.
- Nanoparticles can aid in imaging malignant lesions, so surgeons know where the cancer is, and how to remove it.
- Nanoshells can kill tumor cells selectively, so patients do not suffer terrible side effects from healthy cells being destroyed.
- Dendrimers can sequester drugs to reduce side effects and deliver multiple drugs to maximize therapeutic impact.

Cancer and Nanoshells

Nanoshells are hollow silica spheres covered with gold. Scientists can attach antibodies to their surfaces, enabling the shells to target certain cells, such as cancer cells. In mouse tests, Professor Naomi Halas's research team at Rice University directed infrared radiation through tissue and onto the shells, causing the gold to superheat and destroy tumor cells while leaving healthy ones intact. Technicians can control the amount of heat with the thickness of the gold and the kind of laser. Nanoshells could one day also be filled with drug-containing polymers. Heating them would cause the polymers to release a controlled amount of the drug. Human trials using gold nanoshells are slated to begin in a couple of years.

Cancer and Gold Nanoparticles

Oral cancer is any cancerous tissue growth located in the mouth. Smoking and other tobacco use are associated with 70 percent to 80 percent of oral cancer cases. Thirty thousand Americans will be diagnosed with oral or pharyngeal cancer this year. It will cause over 8,000 deaths, killing roughly 1 person per hour, 24 hours per day.

One oral cancer research group has demonstrated that gold nanoparticles can be bond to malignant cells making cancer easier to diagnose and treat. In laboratory demonstrations, the researchers targeted

Table 5.1 Some Potential Medical Activities Using Nanotechnology during the Next 1 to 5 Years

Cancer	Cardiac Diseases	Diabetes	Infectious Diseases
• Smart probes with reduced toxicity for drug targeting • Identification of biomarkers for early screening in body fluids • Minimal invasive endoscope and catheter for diagnostic and therapy • Implantable mobile systems for detection of cancer cells	• Noninvasive and minimal invasive 3-D imaging techniques for blood flow in the cardiovascular system • Robotics for heart diagnostic and therapy • Telemedicine for heart monitoring using implantable devices	• In vivo diagnostic tools to check glucose metabolism • Minimal invasive glucose sensor and insulin delivery systems for daily home care • Whole body imaging of fat distribution	• Identification of diagnostic markers and easy screening of changes in infected cells. • Integrated diagnostic test for rapid and early diagnosis of viral and bacterial infections

malignant cells in the body with gold nanoparticles. Then they use a laser to heat the particles that destroy the malignant cells.

Breast Cancer and Nanoparticles

Breast cancer is the most common malignancy in women in the United States, with approximately 180,000 new cases diagnosed in this country annually. Breast cancer is the third most common cause of cancer death (after lung cancer and colon cancer) in the United States.

Statistics show that a woman has a one in eight chance of developing breast cancer during her life. In 2007, breast cancer is expected to cause approximately 40,000 deaths.

> *Faster Results for Breast Cancer. Pathologists Use Digital Imaging to Speed up Cancer Diagnosis.* Science Daily. Go to: http://www.sciencedaily.com/videos/2006-02-06/

Some of the treatment of breast cancer includes surgery, hormone therapy, chemotherapy, and radiation therapy. Now scientists are experimenting with nanoparticles to treat breast cancer.

One preliminary plan includes using probes that inject special magnetic iron nanoparticles into a tumor and then heating the nanoparticles. The nanoparticles destroy the cancer cells. In this probe, the magnetic iron nanoparticles, containing antibodies, are concealed in polymers. The polymers make the antibodies nearly invisible to the body's immune system.

> **Did you know?**
> A small percentage of men can also get breast cancer.

The reason for the polymer coating is that you do not want the antibodies to be attacked by the immune system. Inside the bloodstream, the antibodies go to work and attach themselves onto the surface of tumor cells. Then, outside the body, laboratory technicians apply a heat source to the magnetic particles in the tumor region of the body. By applying just the right amount of heat to the tumor region, the heated magnetic particles weaken and destroy the cancer cells.

Please note the use of magnetic particles as a heat treatment to kill breast cancer cells in humans is still conducted only in laboratories at the preclinical and developmental stage. Preclinical tests for humans are still in the future, maybe 5 to 10 years from now.

NANO INTERVIEW: DR. EDITH PEREZ

To learn more about breast cancer, the author was able to contact Edith A., Perez, who is a Medical Doctor and a Professor of Medicine at the Mayo Medical School in Jacksonville, Florida. She is also the Director of Clinical Investigations and the Director of the Breast Cancer Program. Dr. Perez took time out to discuss her professional background and her medical work in breast cancer.

Where did you grow up and what colleges did you attend?
I grew up in Puerto Rico and attended the University of Puerto Rico, the University of Puerto Rico Medical School, and the Loma Linda University for my residency. I also attended the University of California, Davis on a Hematology/Oncology fellowship.

What were some of your favorite activities and subjects in high school?
I enjoyed reading and my two favorite subjects were algebra and chemistry.

What interested you in seeking a career as a medical doctor?
I became interested in medicine after the sudden death of one of my relatives and this added to my goal of getting involved in making a

Dr. Edith A., Perez is a Medical Doctor and a Professor of Medicine at the Mayo Medical School in Jacksonville, Florida. She is also the Director of Clinical Investigations and the Director of the Breast Cancer Program. (*Courtesy Dr. Edith Perez*)

difference for people, more than just dealing with mathematics and numbers.

How did you get involved in breast cancer research?
I identified that there was a significant need for improvements in patient care, which added to my interest in learning, made it a logical career choice. I thought that I could also make a difference in people's lives.

One of your research projects is to continue to collect and evaluate data to demonstrate that adding an anti-HER2 agent (Herceptin) to chemotherapy patients with HER2 positive breast cancer can help prevent recurrence of breast cancer in patients.

What is HER2?
HER2 is a protein found on the membrane (with a portion in the surface) of cells that helps regulate cell growth. However, when the amount and type of HER2 protein are altered, this leads to increased cell growth and possibility of cancer spread, as a result of increased breast cancer aggressiveness.

How does the anti-HER2 agent (Herceptin) treatment work?
Herceptin is a monoclonal antibody that targets the extracellular domain of the HER2 protein. This treatment is used for metastatic breast cancer, but also to diminish the risk of tumor reappearance (and improve survival) after breast cancer surgery. Metastatic, describes a cancer that has spread to distant organs from the original tumor site. Metastatic breast cancer is the most advanced stage of breast cancer. Herceptin slows the growth and spread of tumors that have an overabundance of HER2.

What advice would you give young people who want to explore a career in the medical field and in cancer research?
The medical field, and specifically cancer research, is a great career. There are a lot of facets to be involved with including case and clinical research, patient care, and education. Additionally, the strength of the science and the advances being made and in the horizon makes it an important career for helping society. It is truly an engaging and fulfilling career, both intellectually and on a personal level.

To learn more about Herceptin and Metastatic Breast Cancer Treatment, go to http://www.herceptin.com/herceptin/patient/resources/index.jsp

Cancer and Dendrimers

As you learned in Chapter 2, a dendrimer is a tree-like highly branched polymer molecule. Dendrimers are of particular interest for cancer applications because of their defined and reproducible size, but more importantly, because it is easy to attach a variety of other molecules to the surface of a dendrimer. Such molecules could include antibodies, imaging agents to pinpoint tumors, drug molecules for delivery to a tumor, and molecules that might detect if an anticancer drug is working.

Cancer and Cantilevers

In Chapter 3, you read about the cantilevers attached to the scanning probe microscopes. The cantilever is one tool with potential to aid in cancer diagnosis. Nanoscale cantilevers, tiny bars anchored at one end, can be engineered to bind to molecules associated with cancer. They may bind to altered DNA sequences or proteins that are present in certain types of cancer. When the cancer-associated molecules bind to the cantilevers, there are changes in surface tension that cause the cantilevers to bend. By monitoring whether or not the cantilevers are bent, scientists can tell whether the cancer molecules are present. Scientists hope this bending will be evident even when the altered molecules are present in very low concentrations. This will be useful in detecting early molecular events in the development of cancer.

Cancer and Quantum Dots

Quantum dots also show promises of a new means of diagnosing and combating cancer. Quantum dots, also known as nanocrystals, are miniscule particles, or "dots" made of semiconducting materials. When the dots are stimulated by ultraviolet light, they glow in very intense, bright neon colors.

Emory University scientists have used luminescent quantum dot nanoparticles in living animals to target and image cancerous tumors. The quantum dots were first coated with a protective shell covering. Then special antibodies were attached to the surface of the quantum dots. After the quantum dots were injected into the body, they were guided to a prostate tumor of living mice. Using a mercury lamp, the scientists were able to see the surface of the tumor. It was illuminated by the accumulation of quantum dots on the cell. The scientists believe the ability to both target and image cells in vivo (in the body) represents an important step in the goal to eventually use nanotechnology to target, image, and treat cancer, cardiovascular plaques, and other diseases in humans.

Cervical Cancer and Quantum Dots

Each year, some 10,000 women are diagnosed with cervical cancer, and over 35 percent will die because their cancer was detected too late for treatment to be successful. Easier methods of routinely screening women for precancerous lesions could favor a better survival rate.

Research published in a gynecology journal suggests that the application of quantum dots with other tools, could provide a method for early detection of cervical cancer.

The Targeted Nano-Therapeutics^{TM} (TNT^{TM}) System

The TNT^{TM} System is a noninvasive product that kills cancer using localized lethal heat with negligible damage to healthy tissues. The product is being developed by the Triton Biosystems company and has begun human clinical trials. A component referred to as T-probes are dispensed into the body in serum form by an infusion into the patient's blood stream. Once the T-probes are attached to cancer cells, a focused magnetic field selectively activates the magnetic spheres.

The magnetic energy in the spheres is converted to lethal heat which causes a rapid temperature increase at the surface of the cancer cells, killing them with negligible damage to surrounding healthy tissues. Over a couple of weeks following the infusion, the T-probes completely degrade within the body with no trace or toxicity. The Company believes the TNT^{TM} System will eliminate many of the side effects currently associated with conventional therapies.

DIABETES AND NANOTECHNOLOGY

Glucose is the primary source of energy in the human body. This simple sugar comes from digesting carbohydrates into a chemical that the human body can easily convert to energy. But, when glucose levels in the bloodstream are not properly regulated, a person can develop a serious condition known as diabetes. What happens is that the sugar (glucose) builds up in the blood instead of going to the cells.

The American Diabetes Association estimates that 17 million people in the United States have diabetes, but that one-third are unaware they have the disease. "Diabetes," says the American Diabetes Association, "is a chronic disease that has no cure."

People with diabetes must check their blood sugar levels several times a day to help keep their diabetes under control. For many of the million people diagnosed with diabetes, their daily orders are to watch what they eat and then test their blood glucose levels.

> What are the Symptoms of Diabetes? Who is at Risk? How is Diabetes Confirmed? Diabetes Clinic. IrishHealth. (56K/Dialup. http://www.irishhealth.com/clin/diabetes/video.html#

To control their glucose levels, patients with diabetes must take small blood samples many times a day. Such procedures are uncomfortable and extremely inconvenient. Then, if their glucose level is low, the patient can inject insulin, a hormone that regulates the level of glucose. This is done everyday, with no days off.

Many scientists and researchers are working on ways to solve the diabetes riddle. So a cure, perhaps, is on the way.

Nanorobots and Diabetes

One research group is developing the potential of using nanorobots for treating diabetes. The nanorobots use embedded nanobiosensors in the body to monitor blood glucose levels. The special sensors provide signals to mobile phones carried with the patient. If the glucose level is too low, the nanorobots activate a programmed tune in the cellular phone. The tune alerts the patient to take any necessary action regarding the diabetes. This potential use of the nanorobots may be more convenient and safe for data collection and patient monitoring.

Biosensors for Diabetes

Monitoring blood glucose levels can be tiresome, even with today's improved monitoring devices. Drs. Makarand Paranjape and John Currie, researchers in the Georgetown Advanced Electronics Laboratory (GAEL), are working to take the process to a whole new level.

For the past few years, the team has been developing and testing a new biosensor device for blood glucose monitoring. The biosensor, the size of a small band-aid is designed to be worn anywhere on the body. The biosensor samples tiny amounts of fluids that lie just beneath the skin. The device is small and convenient, and makes measuring glucose levels pain-free and noninvasive. Professor Paranjape and researchers at Georgetown University also have a video showing the procedure called Monitoring Blood Glucose without Pain or Blood.

When you have time, see the video of this biosensor device at: http://college.georgetown.edu/research/molecules/14887.html

See Chapter 7 to read an interview with Professor Paranjape discussing biosensors.

Diabetes Research Continues

Other groups are also developing new tools to treat diabetes. In one study, researchers were able to bond insulin molecules and sugar-sensitive proteins to a biodegradable polymer. The polymer nanoparticles are injected into a repository under the skin. The nanoparticles can detect a diabetic's glucose levels and release appropriate amounts of insulin to keep blood sugar levels steady. The research indicates that diabetics could administer the treatment with only one daily injection, rather than the several pinprick glucose tests and then followed up with insulin shots. An injection a day is all that will be needed. No blood testing. No multiple injections.

Tattoos for Diabetes

Scientists are developing a smart tattoo that could tell diabetics when their glucose levels are dangerously low. Once perfected, the tattoo will allow glucose levels to be monitored around the clock, and could warn the patient if their glucose levels are too low.

The tattoo has been designed by Professors Gerard Cote, of Texas A&M University, and Michael Pishko, of the chemical engineering department at Penn State University. The tattoo is made of polyethylene glycol beads that are coated with fluorescent molecules. The polymer molecules are injected underneath the skin using a needle. Under a light source from a laser or a light-emitting diode, the beads would glow. The amount of fluorescence that is emitted is due to any changes in the concentration of glucose in the bloodstream.

> **Did you know?**
> Most animal cells are 10,000 to 20,000 nanometers in diameter. This means that nanoscale devices, having at least one dimension less than 100 nanometers, can enter cells and the organelles inside them to interact with DNA and proteins.

In other words, the fluorescence levels change according to the amount of glucose present. As an example, a low glucose level would emit a different fluorescent light than a satisfactory glucose level. The fluorescence levels could be measured using a device, such as a watch-like monitor.

IMPLANTS AND PROSTHETICS

In the coming years, patients may receive hip implants and other bone implants that employ nanotechnology. As an example, one research group has created a microsensor for patients recovering from hip replacement and other joint related surgery.

The nanotechnology sensor is permanently implanted with the joint implant. The self-powered wireless microsensor monitors the bone healing process after surgery. During the healing process, the device measures and compares how the bone grows and attaches to the pores on the surface of the implant. It records how the patient is healing and when the patient can resume normal activities.

This technology will not only monitor bone healing at the time of surgery but also can determine when implants are worn out and need to be replaced. The average lifetime of implants is about 15 years. Therefore the microsensor will be valuable throughout the patient's lifetime for observing and maintaining the health of the implant.

NANOTECHNOLOGY AND BURN VICTIMS

Each year, some 45,000 people are hospitalized with burns in the United States. Researchers in the College of Textiles at NC State are creating unconventional "textile" products, such as skin grafts, by harnessing the power of nanotechnology. Russell Gorga, assistant professor in Textile Engineering, Chemistry, and Science at North Carolina State, has been using nanofibers to build a synthetic copy of the body's connective tissue that surrounds and supports cells. The nanofiber skin grafts can be used to regrow the skin of burn or laceration victims. The skin graft provides a porous support structure where cells can attach and nutrients can flow into the cell while waste flows away.

The antibiotic properties of silver have made the metal a popular treatment for wounds and burns. Silver helps skin to heal by preventing infections. In recent years several companies have revolutionized wound care with silver-impregnated dressings that require fewer painful changings.

Another research team has modified using old model inkjet printers to develop a way to make sheets of human skin to be used on burn victims. The printer cartridges are filled with living cells. The printer sprays cell materials, rather than ink, onto a gauze material, creating a sheet of living tissue. The researchers hope that this "skin-printing" method will minimize rejections by the body. This could be a life-saving technology for the 20 percent of burn patients who have the most extensive burns.

DIAGNOSIS AND THERAPY

The main advantage of nanomedicine is to get earlier diagnosis of a disease. Diagnosing a disease early leads to less severe and costly therapeutic demands and an improved clinical care. But the human body is a dark place to observe. So, you need special instruments that can cut through the darkness and provide a window into the body. These kinds of instruments are called imaging tools and you may be familiar with some of them. They include ultrasound, Magnetic Resonance Imaging (MRI), and Positron emission tomography (PET). These instruments allow scientists to see more on everything from observation of individual molecules and their movement within cells.

Molecular Imaging Diagnosis

Nanotechnology can help in designing the next generation of molecular imaging instruments for use in early diagnosis of a disease. The instruments can make it possible to track the effectiveness of pharmaceuticals, treat disease, and monitor the response to therapies. Not only can molecular imaging be used to find, diagnose, and treat disease inside the body, but will also have the ability to show how well a particular treatment is working. As an example, molecular imaging has the potential to suggest how quickly a cancer is growing by determining how rapidly cancer cells are growing and how many are or are not dying. Once this assessment is made, physicians can compile data to determine how best to treat patients with cancers growing at specific rates. Molecular imaging may someday allow discovery of a predisease state in patients.

Lab-on-a-Chip Diagnosis

Laboratory-on-a-chip technology for earlier and quicker diagnosis of a disease is being developed. Lab-on-a-chip has become possible because nanotechnology lets scientists manipulate extremely small materials, on the scale of atoms and molecules. The lab-on-a-chip is a miniaturized, portable version of a blood-count machine. The lab-on-a- chip can be designed for many applications. One example is that it could be a diagnostic tool for cancer detection by searching in blood plasma for certain molecules that could be early indicators of the disease. The handheld device also has advantages since it requires only a little sample of blood to analyze the blood chemistry of a patient. Analyzing the composition of blood is how doctors test for infections and deficiencies in the immune system.

DRUG DELIVERY USING NANOPARTICLES

Researchers believe that the potential benefits of nanotechnology will be to provide vast improvements in drug delivery and drug targeting techniques. These new strategies are often called drug delivery systems (DDS). The goal of a drug delivery system is to deliver the medications to a specific part of the body and to control the time-release rate of the

Lab-on-a-chip. Samir Iqbal, a doctoral student in the School of Electrical and Computer Engineering, works under a laboratory hood to transfer a small amount of DNA solution to electronic chips. DNA attaches to gold molecules on such "biochips" in a technology that offers promise for creating devices for detecting bacteria and other substances by combining proteins, DNA, and other biological molecules with electronic components. Such chips might be used to detect cancer cells and cancer-related proteins. (*Courtesy Purdue News Service photo/David Umberger*)

medication. The drug delivery systems will minimize drug degradation and loss and prevent harmful side effects by delivering therapeutic drugs to the desired site of the body. Drug delivery systems will have potential for many applications, including antitumor therapy, gene therapy, AIDS therapy, and the delivery of antibiotics and vaccines.

Advanced Drug Delivery Systems and Lab-on-a-Chip

Some researchers have discussed the possibility of combining the laboratory-on-a-chip and the drug delivery system. A small device would be implanted into the body, which would continuously monitor the various biochemicals in the bloodstream and in response to an injury or disease would release appropriate drugs. For example, an insulin-dependent diabetic could use such a device to continuously monitor his glucose levels and release insulin into the bloodstream, if needed.

NANOTECHNOLOGY FIGHTS INFECTIONS

Several companies are researching and experimenting with silver nanoparticles that may prevent infections by killing organisms such as *Staphylococcus aureus, E. coli,* and other kinds of infectious bacteria and viruses. In their research laboratories, the companies are using silver (colloidal or ionic silver) or silver in solution as a potent antibacterial product.

Silver is used as an enhancement in making bandages, dressings, and other health-care products. The silver nanoparticles or nanocrystals kill bacteria, viruses, and germs. The technicians break down silver into nanoparticles that are much smaller than naturally occurring particles. This process creates greater solubility and an improved ability for the silver atoms to migrate through a germ's cell walls to kill the organism.

The antibacterial and antiviral properties of silver are found in the products of other industries as well. One shoe manufacturer is using silver fiber built into shoes to prevent foot fungus and bacteria from growing. This is important to diabetics, as they are prone to serious foot infections such as gangrene, which can result in amputation of the foot or limb if infection is allowed to set in.

Several manufacturers of outdoor equipment are also using silver fiber in their products to prevent bacteria and fungi from growing in their sleeping bags, cold-weather jackets, hiking clothes, and cycling jerseys.

PHARMACEUTICAL NANOTECHNOLOGY RESEARCH

The National Science Foundation (NSF) expects nanotechnology to account for around half of all drugs made and sold by 2010. Demand for nanotechnology health-care products in the United States is projected to increase nearly 50 percent per year to $6.5 billion in 2009. There will be the potential of introducing new, improved cancer and central nervous system therapies. Diagnostic tests, based on quantum dots and imaging agents based on iron oxide nanoparticles will also show strong growth, according to pharmaceutical economists. By 2020, demand for nanotechnology health-care products is projected to exceed $100 billion.

The need for new or improved medicines in several therapeutic areas will lead to the increasing use of nanotechnology in pharmaceutical applications. These areas include protein-based compounds for cancer, diabetes, and for infectious diseases. Experts also predict that over the

Sarah Perrone and Matt Boyas are high school students who attend Upper St. Clair High School in Upper St. Clair, Pennsylvania. Matt and Sarah entered the ExploraVision Contest in 2006. Their project was called NAI: Nanotech Active Immunization. (*Courtesy Elise Boyas*)

long term, pharmaceutical applications for nanotechnology will extend into mostly therapeutic areas and in drug delivery systems.

NANO INTERVIEW: MATT BOYAS AND SARAH PERRONE

This chapter documents an interview with two students, Matt Boyas and Sarah Perrone. They received a 2006 NSTA/Toshiba ExploraVision award for their project entitled *NAI: Nanotech Active Immunization* which used nanotechnology. The NSTA/Toshiba ExploraVision competition is held annually and encourages students to think of possible future uses of technology. Matt and Sarah attend the Upper St. Clair School District, located outside of Pittsburgh, Pennsylvania.

Their teacher, sponsor, and coach is Ms. Patricia Palazzolo. She is the Gifted Education Coordinator, grades 7–12, for the Upper St. Clair School District in Upper St. Clair, Pennsylvania. In addition to finding hands-on "real world" learning opportunities for her students, Ms. Palazzolo also coaches Future Problem Solving teams and teaches a special course on the impact of technology on every aspect of society. (The Future Problem Solving program's Web site is www.fpsp.org)

What is the Toshiba/NSTA ExploraVision Awards Program?
Before we proceed to the interview, Patricia Palazzolo was kind enough to fill me in on the steps that are needed to submit a project for the Toshiba/NSTA ExploraVision Awards Program. These are her words:

"The students are asked to produce an 11-page paper and a 5-page design for a related Web site based on their ideas for the future of any current technology or technological item. They do a tremendous amount of research to learn about the beginnings and cutting-edge of the technological item. They then use their creative and critical thinking skills to imagine what that item might be like 20 years from now, what breakthroughs would be necessary to turn their futuristic idea into reality, and what the ramifications—both positive and negative—would be if their idea were to become reality. There are no three-dimensional models required for the initial submission, although students reaching the higher levels of the competition are required to create a visual representation of a prototype of their invention."

See Chapter 10 for more information about the Toshiba/NSTA ExploraVision Awards Program. *The Web site is www.exploravision.org*

Nano Interview. Matt Boyas and Sarah Perrone, NAI: Nanotech Active Immunization
"Using millions of nanobots, NAI seeks to create a world in which people never get a virus."

Entered in the Toshiba ExploraVision Contest in 2006, NAI is the work of *Matt Boyas* and *Sarah Perrone*, students at Upper St. Clair High School in Upper St. Clair, Pennsylvania, a Pittsburgh suburb. We had a chance to sit down with Matt and Sarah and ask them a few questions.

What was the inspiration for NAI?
Matt: At the time we were beginning the project, bird flu was a concern for many people of the world. We decided to try and come up with a way to eliminate the fear of the virus.

Sarah: However, we did not want to make our invention limited to a certain time period or a certain ailment. We then broadened our look and came up with NAI.

Before you could invent NAI, you had to research current immunization methods. What are some?
Sarah: The main precursor to NAI is the vaccine. The traditional shot-based vaccine is the main method of immunization in our world today, but scientists are coming up with new methods including mist and cream vaccines.

What exactly is NAI?
Matt: NAI: Nanotech Active Immunization is a plan to successfully immunize the masses without the dangerous side effects of needle-based vaccinations. NAI uses nanotech robots called GNATs (Gargantuan Nanotech Antigen-removal Team) to enter the body and destroy the viral genetic material using enzymes.

How are the GNAT robots structured?
Sarah: The GNAT robot is a tiny, self-controlled robot. The robots are airborne, and are inhaled into the body. These robots float around in the bloodstream until a virus is detected, whereupon the robots are activated and destroy the virus. Each GNAT robot is powered by thousands of muscle cells growing on the silicon chips. The cells are very much alive, and produce ATP during aerobic respiration, which powers them.

How do the GNAT robots eliminate viral infection?
Matt: Each GNAT robot has sensors that detect the protein capsid, a component of all viruses. Once detected, the robot rushes to the virus, where it inserts a long needle into the nuclear matrix. DNase and RNase enzymes, made by pancreatic cells on the robot, travel through the needle and break down the genetic material in the virus. Without any genetic material, the virus cannot infect the person. White blood cells destroy the "empty" virus, and the robot goes into StandBy mode until another virus is detected.

What are some anticipated benefits or risks to using NAI?
Sarah: Well, the benefits of NAI are obvious. No one would get infected by viruses anymore. This would remove the threats of thousands of different viruses, from the common cold to HIV.

Matt: However, NAI, as with all medications, may have its risks. Since we haven't really made it, we can't say what risks it may have, but we can put out some guesses. Since the GNAT robots are located in the air, the robots may affect children or animals differently than adults. Also, there is always the possibility of a technology malfunction.

Are there any institutions that are currently using nanotechnology in a similar way that you propose to use it in NAI?
Sarah: NAI is a completely theoretical project, but modern science is well on its way to making it a reality. A needle-free, nanotechnology-based vaccine has been developed for anthrax. In addition, similar nanotechnology-based vaccines have been developed for bubonic plague, staph, and ricin for the U.S. Army, as well as one for malaria for the U.S. Navy.

READING MATERIALS

Freitas, Jr., Robert A. *Nanomedicine: Basic Capabilities, Vol. 1.* Austin, TX: Landes Bioscience, 1999.

Goodsell, D.S. *BioNanotechnology: Lessons from Nature.* Hoboken, NJ: Wiley-Liss, Inc., 2004.

Jones, R. L. *Soft Machines: Nanotechnology and Life.* Oxford, UK: Oxford University Press, 2004.

Rietman, Edward A. *Molecular Engineering of Nanosystems Series: Biological and Medical Physics, Biomedical Engineering.* New York: Springer, 2001.

Scientific American (author). *Understanding Nanotechnology.* New York: Grand Central Publishing, 2002.

VIDEOS AND AUDIOS

Nanobumps Improve Joint Replacements. Museum of Science, Boston

Research from Brown University shows that adding nanoscale texture to the surface of orthopedic implants not only helps prevent infection, but it also helps the implant join to a patient's natural bone. http://www.mos.org/events_activities/videocasts&d=1185

Bones that Grow Back Video: http://www.sciencentral.com/

Dendritic Polymer Adhesives for Corneal Wound Repair

Presented by: Mark W. Grinstaff, PhD., Metcalf Center for Science and Engineering: http://www.blueskybroadcast.com/Client/ARVO/

National Cancer Institute: Video Journey into Nanotechnology: http://nano.cancer.gov/resource_center/video_journey_qt-low.asp

Audios

Voyage of the Nano-Surgeons. NASA-funded scientists are crafting microscopic vessels that can venture into the human body and repair problems—one cell at a time. http://science.nasa.gov/headlines/y2002/15jan_nano.htm

Next Generation of Drug Delivery. The Bourne Report.

There are all sorts of ways to get medicine into the body; here are a few examples of how MEMS and Nanotech-based approaches are making a difference. Marlene Bourne. http://bournereport.podomatic.com/entry/2006-12-10T13_46_12-08_00

WEB SITES

Foresight Insitute, Nanomedicine: Nanomedicine may be defined as the monitoring, repair, construction and control of human biological systems at the molecular level, using engineered nanodevices and nanostructures. http://www.foresight.org/Nanomedicine

National Heart, Lung, and Blood Institute: http://www.nhlbi.nih.gov/index.htm

National Cancer Institute: http://www.cancer.gov/

Scientific American: http://www.sciam.com/nanotech/

Nano Science & Technology Institute: http://nsti.org

The National Institute for Occupational Safety and Health: http://www.cdc.gov/niosh/homepage.html

SOMETHING TO DO

Cancer Microscope Slide Activity. You may be interested in viewing human cancers and normal human tissues prepared on special microscope slides. The VWR International, a large distributor of science supplies for classrooms, offers several sets of cancer slides. One set is called, *Array 1: Top 4 Human Cancers and 4 Normal Human Tissues.* If interested, you can contact them at www.vwr.com You can also check other school science suppliers and your school science department to find if they have these kinds of slides.

6

The Business of Nanotechnology

NANOTECHNOLOGIES IN BUSINESSES

The National Science Foundation predicts that the global marketplace for goods and services using nanotechnologies will grow to $1 trillion by 2015. The United States invests approximately $3 billion annually in nanotechnology research and development. This dollar figure accounts for approximately one-third of the total public and private sector investments worldwide.

The range of possibilities of nanotechnology-manufactured products—from electronics to communications, aerospace, medicine, energy, construction, and consumer goods—is almost limitless. More than one-half of the major corporations that are in the stock market are in the nanotechnology business now or will be in the future. See the Appendix for a listing of companies that are in the nanotechnology field.

Last year, more than $32 billion in products with nanotech materials were sold worldwide. By 2014, some $2.6 trillion in manufactured goods could contain nanotechnologies, according to a research group that tracks the industry.

The researchers also estimated that about a little over 1,000 companies have claimed to be working in a field that is related in some way to nanotechnology. In addition, the researchers also indicated that there are 1,500 companies that are exploring nanotechnology options.

SPORTING GOODS EQUIPMENT

The sporting goods companies have been adapting nanotechnology for several years. In fact, they may have been the earliest of commercial companies to use nanotechnology applications. As an example, some hockey equipment manufacturers have developed carbon nanotube composite hockey sticks that are more durable and have better impact resistance than traditional sticks.

Baseball Bats

One sports manufacturer has created a special bat they call the CNT. The "CNT" stands for carbon nanotube technology. In their traditional carbon fiber bats, the spaces inside contain only resin. Over time the resin can weaken the bat's power. The company solved the problem by applying a special carbon nanotube solution into the resin. The result is that the bat gives hitters more power through the hitting zone. These new bats can cost about $175 and up.

Tennis Rackets

New kinds of nanotech tennis rackets are now in sporting goods retail stores around the world. One of the latest tennis racket features a small proportion of nanotubes that are located in the yoke of the racket. This is the part of the racket that tends to bend slightly with the impact of a hard-hitting tennis ball. According to tests, the nano tennis rackets bend less during ball impact than the traditional rackets. The rackets, which weigh 245 to 255 grams cost about $230.

Nano Golf Balls

Wilson announced it was the first golf equipment manufacturer to strategically use nanotechnology. The application of the cutting-edge technology has allowed Wilson Sporting Goods Company to develop stronger and lighter materials that optimize the performance of their lineup of sporting goods equipment.

Not to be left out, other golf equipment companies have also figured out how to alter the materials in a golf ball at the molecular level so the weight inside shifts less as the ball spins. This means that even a badly hit ball goes straighter than other kinds, according to the manufacturer.

> *Golf Gear Improving Your Score with Nanotechnology.* AOL Video. http://video.aol.com/video-search/Golf-GearImproving-your-score-with-technology/id/3305225634

Skis and Nanofibers

Every skier wants to improve his or her performance on the slopes. Ordinary skis may have tiny voids that create stress points that weaken the ski. One company now injects nanoparticle crystals into these voids.

Table 6.1 Some Examples of Nanotechnology Products

Consumer Products	Medical Supplies	Components	Equipment
Sports Equipment	Burn and Wound Dressings	Automobile and Aircraft Parts	Lithography Equipment
Video Games	Drugs	Catalytic Converters	Scanning Electron Microscopes
Stain-Free Clothing	Diagnostic Tools	Fuel Cells	Atomic Force Microscopes
Cameras	Medical Imaging Equipment	Solar Cells	Nanotweezers
Sunscreens and Cosmetics	Artificial Muscles	Flat Screens	Biosensors
Paints	Dialysis Filters	Transistors	Cantilevers
Food Packaging Products		Batteries	Computer Aid Design
		Toxic Wastes Cleanup Equipment	Air Purifiers
			Super Capacitors
			Water Filtration Systems

The result is that the skis offer a low swing weight, easy turning, and durability.

Nanotechnology applications will continue to make better-performing sporting goods equipment. Part of the potential nano list of equipment includes yacht racing masts, vaulting poles, softball bats, golf clubs, and making lighter racing bikes and Indianapolis racing sport cars.

CHEWING GUM AND NANOCRYSTALS

Did you ever wonder why there is no chocolate chewing gum? One reason is that the cocoa butter in the chocolate could not mix well with the polymers, the chemicals that give gum its elasticity. In fact, the fats found in chocolate will cause chewing gum to fall apart. A Chicago-based company, called O'Lala's, developed a solution incorporating nanoscale crystals. The crystals give its gum a creamier texture and chocolate flavor. A pack of 12 pieces will cost you about $1.25.

APPAREL INDUSTRY

Several clothing companies have come out with a new brand of non-stain fabric for pants that uses nanotechnology. The fabric resists spills from many types of fibers such as cotton, synthetics, wool, silk, rayon, and polypropylene. The nanomade fabric also repels a range of liquids including beverages and salad dressings. The fabric keeps the body cool and comfortable and has an antistatic treatment that reduces static cling from dog hair, lint, and dust.

Nanotechnology in Cleaner Clothes

David Soane, a chemical engineer, started a company called Nano-Tex. He uses the principles of nanotechnology to improve the strength and durability of natural fibers like cotton. He created tiny structures that he calls "nanowhiskers," which are tiny hairs that make liquid spills bead up and roll right off various fabrics.

Soane discovered his idea for nanowhiskers by washing a peach, a fuzzy kind of fruit. "When you wash a peach, very often the water rolls right off," explains Soane. "That's because on the fruit's surface, there are all these little pointed whiskers." The nanowhiskers can repel stains because they form a cushion of air around each cotton fiber. When something is spilled on the surface of the fabric, "the miniature whiskers actually prop up the liquid drops, allowing the liquid drops to roll off," says Soane, who calls his stainproofing process, Nano-Care.

> Cleaner Kids Clothes. A Nanotechnology Fabric. ScienCentralNews. http://www.sciencentral.com/articles/view.php3?article_id=218392126&language=english

Each of Soane's synthetic nanowhiskers is only 10 nanometers long, made of only a few atoms of carbon. "They repel a range of fluids," says Soane, "including coffee, tea, salad oil, ketchup, soy sauce, cranberry juice."

Nanotechnology Socks

Another nano-improved apparel application that is selling in United States Military Stores, stateside and abroad, are specially-made polyester socks. The company ARC Outdoors uses a special process that allows nanometer silver particles to bond within the fibers of the sock. The silver has antimicrobial properties that provide protection against odor and fungus in socks.

COSMETICS

Several cosmetic companies have already marketed a number of products that include nanotechnology applications. These products include: deodorants, antiaging creams, and sunscreens.

Sunscreens and Skin Cancer

Skin cancers are caused by abnormal cell changes in the outer layer of skin called the epidermis. Skin cancer is the most common cancer in the world. Most cases of skin cancer can be cured, but the disease is a major health problem because it affects so many people.

Excessive exposure to sunlight is the main cause of skin cancer. Sunlight contains ultraviolet (UV) rays that can alter the genetic material in skin cells causing mutations.

Doctors believe that if you are at risk for skin cancer, you should follow certain precautions that include avoiding intense sun exposure from late morning through early afternoon. While outdoors, wear a hat, long sleeves, trousers, and sunglasses that block UV radiation. Another precaution tip is to use a sunscreen, a sun blocker, with a sun protection factor of 15 or higher whenever you are outside.

Zinc is one of the most effective sun blockers used in sunscreens because they can absorb much of the harmful ultraviolet (UV) rays that can damage human skin. However, because of the large particles of zinc, the sunscreen looks thick and leaves white marks on the skin. The large particles of zinc both absorb and scatter light, which is why zinc cream looks white. But if you make the zinc particles small enough, the sunscreen will appear transparent because it will let visible light pass through.

> **Did you know?**
> Ultraviolet rays can cause skin cancer. According to medical reports, the number of people susceptible to skin cancer is predictably higher in places with intense sunshine, such as Arizona and Hawaii. Skin cancer is most common in Australia, which was settled largely by fair-skinned people of Irish and English descent.

So, researchers at the cosmetic companies went to work to reduce the large zinc particles into ultrafine nanoparticles. Their reduction of the size of the zinc particles is impressive. The large zinc oxide particles were reduced from about 250 nanometers to as tiny as 10 nanometers.

These ultra fine nanoparticles are especially effective in absorbing light in the damaging ultraviolet range, below 400 nanometers wavelength. The evidence so far is that a sunscreen containing an equivalent

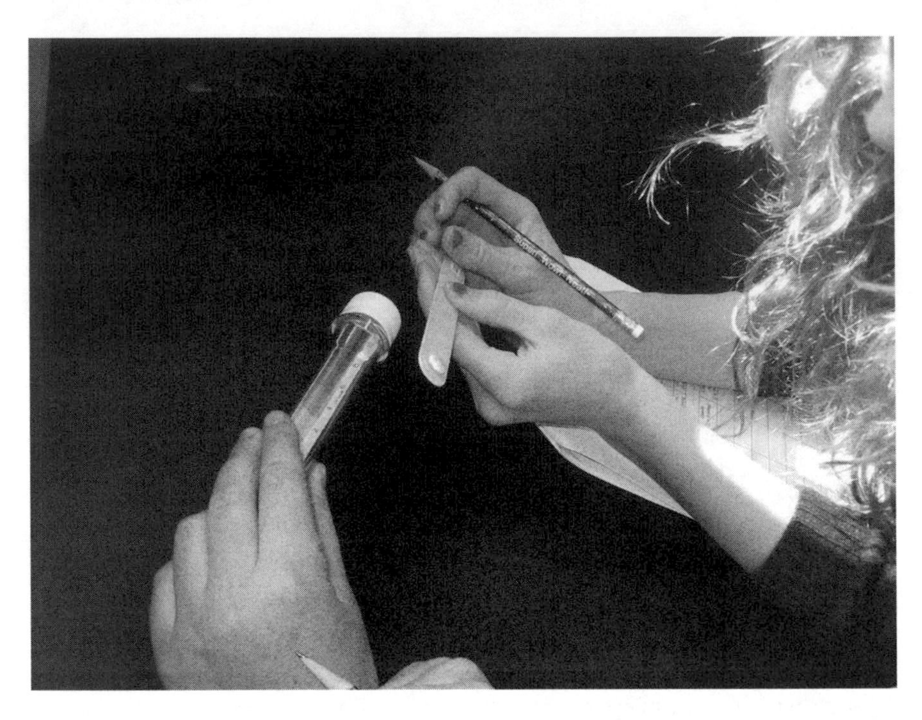

A student in the NanoSense project is testing the UV-blocking ability of a sunscreen. She is observing and recording what happens to the skin when the UV light reaches it. To learn more about this activity, refer to the instructions at the end of this chapter for more information. (*Courtesy Patricia Shank*)

amount of ultrafine particles could be about twice as protective against UV radiation as a normal sunscreen. However, regulation of these new kinds of sunscreen products will be needed because of the limited research that has been carried out in this field.

Nano Sunscreens. To see how nano sunscreens work go to: http://online.nanopolis.net/viewer.php?subject_id=274

APPLIANCES

Nano Silver Seal Refrigerator

One major kitchen appliance company has offered a new line of refrigerators and washing machines that uses nanocoatings to prevent germs and molds from growing inside the appliances. Silver, as small as 1 nanometer across, is used to coat surfaces. These nanoparticles are so electrically active that they inhibit the growth of harmful bacteria and fungus. In a refrigerator, one company used the nanosilver in the

deodorizer unit and in the water dispenser to sanitize the air and water that passes over them.
[Note. The Environmental Protection Agency is investigating to find if silver particles can be harmful when they are applied in the manufacturing of some appliances.]

Flat-Plane Display Screens

Since 2003 several large electronic companies began producing the latest flat-screen display technology for laptops, cell phones, digital cameras, and other uses that are made of nanostructured polymer films. These special displays are known as OLEDs, or organic light-emitting diodes.

These ultrathin displays are manufactured by sandwiching extremely thin (often nano-sized) layers of organic polymer light-emitting materials between electrodes. The images are bright and viewable at wide-angles. The displays are smaller and lighter-weight than traditional LCD displays—meaning they are ideally suited to mobile electronics, such as digital cameras, cellular phones, and handheld computers. Among OLED screen advantages are brighter images, lighter weight, less power consumption, and wider viewing angles.

The Business of Air Purifiers

Everyone wants clean air indoors. In fact, indoor pollution may be more of a problem in many areas than outdoor air pollution. So companies are building better indoor air purifiers. One company, NanoTwin Technologies, has patented NanoTube as the active component behind their NanoBreeze Air Purifier. The tube is wrapped in a fiberglass mesh coated by a layer of titanium dioxide crystals that measure 40 nanometers in size. When switched on, the NanoTube radiates UV light that charges the crystals to create powerful oxidizing agents that destroy airborne germs and pollutants circulating over the tube's surface.

ELECTRONICS AND COMPUTERS

The best computer chips used presently can hold around 40 million transistors, each measuring around a micrometer (one-thousandth of a millimeter) or less. This is electronics on a microscale, and it is done using what is called a top-down approach. It involves taking material, usually silicon, and slicing it up into many wafers. Patterns (circuits consisting of many transistors) are drawn on light-sensitive films on the surface of each wafer. Using a light source, the unwanted material between the circuits is etched away with acid. The result is a finished product. This manufacturing approach is called top down because you start with something large, and then cut it down to a product you want. Most traditional chip manufacturers use top-down processes.

However, nanotechnology researchers are also looking at another completely new way of manufacturing transistors. The new way requires a bottom-up approach. In this process, molecules are created on the surface of a chip allowing the molecules to self-assemble into larger structures used to make transistors.

AUTOMOBILE/VEHICLE INDUSTRY

Nanomaterials could make future automobiles and airplanes lighter and stronger and most of all improve fuel efficiency. Nanotechnology promises to make everything from new coatings that are scratch-resistant and self-cleaning to batteries that last longer for hybrid vehicles. In the longer term, nanotechnology applications will most likely make hydrogen easier to produce and safer to store as a renewable energy source for vehicles. Most experts agree that nanotechnology will play an important role in advancing hydrogen fuel cell technology. See Chapter 8 for more information about hydrogen fuel cells.

Auto Catalytic Converters

Automakers are developing new ways to improve on their catalytic converters that are used to control automobile emissions.

Today conventional technologies utilize precious group metals called PGMs. The PGMs, such as platinum, palladium, and rhodium, are the main catalysts used in catalytic converters to control automobile emissions. However, the PGM demand has resulted in higher platinum prices. Also, the need for more PGMs is expected to continue to rise with

- an increase of more diesel cars in Europe,
- higher emission standards for trucks in North America and Japan, and
- an increase of vehicle production and higher emission standards in China and India.

Several nanofabrication companies are now developing nanocomposites catalysts that will use much lower levels of PGMs in catalytic converters.

Table 6.2 Nanotechnology and the Automobile

Auto Frames	Exhaust System	Engine	Paint and Tires
Carbon nanotube alloys will be used to provide lightweight materials that will provide both strength and lightweight materials for frames, car doors, bumpers, mirrors, and windshields.	Nanoscale metal oxide catalysts will reduce harmful emissions.	Nanocatalysts and membrane technology will play a major role in using fuel cells that will be more economical and cleaner than the present internal combustion engine.	Nanopowders and special coatings will increase the durability of paint coatings. Nanocomposites will be used to build tires that will improve skid resistance and abrasion.

Automobile Paint and Waxes

A new type of finish for automobile bodies offers improved scratch resistance compared to conventional paint finishes. The nanoparticle clear coat lacquer fills in and conceals scratches, which provides a nice glossy coat to the look of the car or truck.

Several nano car waxes are now on the market and they are made with nano-sized polishing agents. The new waxes provide a better shine due to its ability to fill-in tiny blemishes in automotive paint finishes.

The automobile companies are also spending money on nanotechnology to manufacture better-built vehicles to make them lighter, stronger, safer, and more fuel-efficient. For strength and safety, they are using nanocomposites in the frames, doors, engines, seats, tires, and in other auto parts as well. They are experimenting and testing hydrogen fuel cells to power their vehicles, making them less dependent on petroleum-based fuels and better for the environment.

Aeronautic companies are researching ways to use nanosensors that could be built directly into the body of an aircraft and used to identify mechanical, structural, and electrical problems before they ever arise. Among the more innovative things nanotechnologists envision are self-healing materials and shape-shifting wings.

AIRCRAFT POTENTIAL AND METAL RUBBER

In Chapter 2, you read about Dr. Claus, the President of NanoSonic Inc. that makes Metal Rubber™. As you recall, Metal Rubber™ looks like brown plastic wrap and has some amazing properties, including elasticity and it is very tough. Dr. Claus said that one of the most exciting applications of Metal Rubber™ is to use it for "morphing aircraft structures. These are aircraft that dynamically change the shape of their wings and their control surfaces during flight," he explains. Almost the way that a hawk might fly along, see prey, and change its shape to dive down. The hawk changes the shape of its body, and when it does that, it needs to be able to sense what the outside forces and pressures are so it knows how to fly. For a plane, you need a material that's mechanically flexible. But you also need a material with a surface that's controlled by sensors and electrical conductors that allow it to do that sensing and change shape accordingly. Metal Rubber™ might allow sensors that can be flexed. Now NanoSonic is working in partnership with an aircraft company to explore Metal Rubber™ potential in aerospace.

PAINT AND OTHER WATER RESISTANCE COATINGS

Another example of rapid growth in nanotechnology applications is in the field of wear-resistant paints and other kinds of coatings. Beginning in 1996, the Department of Defense (DOD) supported funding for developing processes to manufacture coatings for use in the marine environment.

Then in 2000, the first nanostructured coating was qualified for use on gears of air-conditioning units for U.S. Navy ships. DOD estimates that use of the coatings on this equipment will result in a $20 million reduction in maintenance costs over 10 years. The development of wear-resistant coatings by the DOD will lead to commercial applications that can extend the lifetime of moving parts in everything from personal cars to heavy industrial machinery.

Paints and Anti-Graffiti Coating

Graffiti is an expensive social concern for many major cities. Two examples illustrate the scale of the problem: the subway system or the tube in London spends over 15 million dollars per year for cleaning up graffiti and the City of Los Angeles spends even more millions for cleanup work.

Several companies have used nanotechnology to create paints and coatings to help solve the graffiti problem. Some of the products, so far being developed, provide a permanent clear coating that does not allow paint, permanent markers, stickers, graffiti, or any other markings to bond to the coating. The clear coating forms a barrier between the surface of, say, a wall and any graffiti paint that might be applied.

REMOVING WINDSHIELD FOG

Fog conditions can be a major concern for drivers in many parts of the world. Many vehicle accidents have occurred due to fogging conditions that have occurred rapidly. One company may have an answer to keep the fog particles from forming on the windshields of cars and trucks. They are developing a way to heat up a windshield without the use of costly copper heating elements. The group has designed a special coating, consisting of a transparent coat of carbon nanotubes (CNT), to cover the surface of a windshield. When electricity is supplied to the windshield, it becomes a heater that can clear the surface of a windshield in a short period of time.

SELF-CLEANING GLASS

Tired of cleaning windows? After years of development, Pilkington Activ™ is the world's first self-cleaning glass to use a microscopic coating with a unique dual action. The coating on the glass reacts with ultraviolet light given off by the sun. This reaction on the surface breaks down and loosens the organic dirt. Then, when it rains, the coating causes rain water to "sheet" off the surface of the glass, which not only washes away the loosened particles of dirt, but also prevents the formulation of droplets, which cause streaks and make windows look dirty.

ANTIBACTERIAL CLEANSERS

There are several antibacterial cleansers that use nanoemulsion technology to kill pathogens. These cleansers are able to kill tuberculosis and bacteria while remaining nonflammable, noncorrosive, and nontoxic. The good news is that there are no harmful effects when using these products.

MEDICAL BANDAGES

Silver's antibiotic properties have made the metal a popular treatment for wounds and burns. Special dressings for burns offer antimicrobial barrier protection using concentrations of silver with nanocrystalline technology. It helps skin to heal by preventing infections during treatment. The silver-impregnated dressings require fewer painful changings than previous silver treatments.

SOLAR ENERGY: PHOTOVOLTAIC CELLS

In this century, the use of solar cells is a major goal for supplying energy needs for the United States as well as the world. As of 2005, the world market for solar cells was about 3 to 4 billion dollars. Of this amount Japan had a 20 percent share, according to economic experts. Germany had the greatest number of solar installations, accounting for 57 percent. The United States accounts for 7 percent and Europe accounts for 6 percent. The rest of the world accounts for 10 percent.

Many companies are focusing on the technology in the production of solar cells. However, the technology still remains expensive when compared to the costs of fossil fuels. Presently solar cells have two major problems: they cost too much to make (in the form of energy), and they are not very efficient. Typical efficiencies for solar cells range from 10 to 15 percent.

Scientists have been doing a lot of research and experiments with quantum dots to make photovoltaic cells more efficient. A new quantum-dot-based solar cell has recently been prototyped with 30 percent efficiency.

One company, Nanosolar, has developed a low-cost technology to make solar cells based on the economics of printing. Their technology is like printing solar film on paper. To make the thin-film solar cells, Nanosolar prints CIGS (copper-indium-gallium-selenium) onto a thin polymer using machines that look like printing presses. There is no costly silicon involved in the process. According to the company, a solar cell from Nanosolar will cost about one-fifth to one-tenth the cost of a standard silicon solar cell.

Did you know?

CIGS are made up of four elements. Copper (Cu) is a reddish metal that is used mostly for electrical equipment. Indium (In) is a soft, metallic metal that is used as a protective plate for bearings and other metal surfaces. Gallium (Ga) is a soft blue-gray metal. As gallium arsendite (GaAs), it is used in light-emitting diodes (LEDs). Selenium (Se) is a nonmetallic chemical element used extensively in electronics, such as in photocells.

The technology of producing thin-film solar cells has the potential to achieve mass production at low costs. One economic researcher estimates that the market for thin-film solar cells will increase in the next few years, and then reach $1.5 billion in sales in 2012. See Chapter 8 for more information about photovoltaic cells.

BATTERY TECHNOLOGY

Lithium-ion batteries, commonly used in laptops and cell phones, have a relatively short life span because they use a graphite anode, which wears out quickly with normal usage. Altair Nanotechnologies has developed nanomaterials that resulted in safer, more powerful batteries.

The Altair team substituted the carbon anode with lithium titanate (LiTiO) nanocrystals, which makes the battery last longer. Lithium Titanate or Lithium Titanium Oxide nanocrystals are typically 10 to 100 nanometers.

Using the lithium titanate nanocrystals also lengthened the life of a lithium-ion battery from 750 recharges to between 10,000 and 15,000 recharges. The nanocrystals also made the battery safer.

Did you know?
Lithium (Li) is the lightest metal and a thin layer of it can float on water. It is a strong alloy used in building aircraft and space vehicles.

Virus-Assembled Batteries

Professor Angela Belcher, a biomolecular materials chemist at MIT, is trying to use biological methods, such as viruses, to assemble batteries. "The goal is to have biology make things in an environmentally friendly way," says Professor Belcher.

Professor Belcher's virus-assembled batteries are thin, transparent sheets that look like plastic wrap. They could be used to create smart credit cards or lightweight hearing-aid batteries. Eventually, Professor Belcher hopes to weave battery cells into textile fibers to create battery-powered fabrics for clothing. As an example, soldiers might plug night-vision goggles into their uniforms, instead of using traditional batteries for energy.

Long-Lasting Batteries

The company mPhase Technologies and Bell Labs, research and development part of Lucent Technologies, have teamed up to develop a type of nanobattery. The nanobattery can store and generate electric current that could be used for many years after it is bought.

In traditional batteries, all the chemicals are mixed together in the battery. This causes traditional batteries to degrade even when not in use. Therefore, traditional batteries have a certain shelf life before they expire. In the new nanobattery, the chemicals are not mixed together, until you activate it. In other words, no power is used until the device, such as a tool, is turned on. So, the battery does not dissipate any of its chemical energy when not in use. This feature provides a very, very long shelf life that regular batteries don't have.

THE BUSINESS OF BUILDING ATOMIC FORCE MICROSCOPES

Several companies such as Asylum Research, Veeco, Agilent, Novascan specialize in sales and services of Atomic Force Microscopy (AFM). The AFM is a high-resolution imaging and measurement tool that allows researchers to directly view single atoms or molecules that are only a few nanometers in size. Then it produces a three-dimensional map of the sample's surface.

Atomic force microscopes are a significant portion of the $1 billion market for nanotechnology measurement tools. The price of an AFM can cost up to $500,000.

Atomic force microscopy (AFM) is the principal technology that scientists and researchers use to view and manipulate samples at the nanometer scale, which is why it's called the "eyes of nanotechnology."

The AFMs are employed in the semiconductor industry, the materials science field, and the biotechnology industry. AFM life science applications include the imaging of live cells, proteins, and DNA under physiological conditions; single molecule recognition; and the detection of single biomolecular binding interactions.

The AFM has several advantages over other technologies, which makes it a favorite with researchers. The chief difference between AFM and other microscopy techniques is the measure of resolution. While electron and optical microscopes provide a standard two-dimensional horizontal view of a sample's surface, AFM also provides a vertical view. The resulting images show the topography of a sample's surface. While electron microscopes work in a vacuum, most AFM modes work in ambient or liquid environments. AFM does not require any special sample preparation that could damage the sample or prevent its reuse.

The number of applications for AFM has increased since the technology of this kind of microscope was invented in the 1980s and now spans many areas of nanoscience and nanotechnology. AFM provides the ability to view and understand events as they occur at the molecular level. This will increase knowledge of how systems work and lead to advancements in such areas as drug discovery, life science,

The SRI International NanoSense Project team includes (left to right): Anders Rosenquist, Patricia Schank, Alyssa Wise, Tina Stanford. The team designs activities so high school students can understand the science concepts that account for the nanoscale phenomena. (*Courtesy Larry Woolf*)

materials science, electrochemistry, polymer science, biophysics, and biotechnology.

NANO INTERVIEW: DR. ALYSSA WISE AND DR. PATRICIA SCHANK (NANOSENSE TEAM) AND DR. BRENT MACQUEEN (NANOSCIENTIST)

A team of educators and scientists at SRI International in Menlo Park, California, have designed and implemented a project called NanoSense. NanoSense introduces high school students and teachers to nanoscale science concepts and skills. The instructional materials in the program include hands-on activities, student worksheets, lesson plans, visualizations, and student readings. Hundreds of high school students have used NanoSense materials.

[Note: At the end of this chapter you will have the opportunity to do a nano activity, *Sunscreens and Sunlight Animation,* from one of the NanoSense units.]

I invited the team at SRI International to reply to questions about nanoscience and nanotechnology asked by the high school students who participated in the NanoSense program. The following are a few of these questions.

Where does most nanoscience research take place?

Nanotechnology is a big business. More than $8.6 billion was invested by governments, companies and venture capitalists worldwide in 2004 and $60 billion to $70 billion worth of products that incorporate nanotechnology are sold annually in the United States. The National Nanotechnology Initiative (NNI) estimates that more than two million people will be working in nanotechnology within the next 10 years.

The United States is the leader in nanotechnology research, but not by much. Nanotechnology is the largest federally funded science initiative in the country, receiving over $1 billion in federal funds annually, and private companies invest at least another $1 billion. Initially, governments made the largest investments in nanotechnology, but this has changed in recent years, with industry contributing larger and larger amounts.

Japan has been involved in nanotechnology since the very early stages and is very close to the United States in its level of development; the Japanese government invested around $750 million in nanotechnology in 2004. The Europeans have recently recognized the importance of nanotechnology and are starting to put a lot of money into nanotechnology projects as well. Finally, countries that are beginning to get into nanotechnology include India, China, South Africa, and Brazil.

How are nanosized particles made?

Typically we start from the bottom and build up the particles. You begin by reacting the molecules to form a mixture in which solid particles are suspended in a liquid. Then you pour this mixture into a mold, where it becomes a soft, jelly-like material. Finally, you carefully apply heat and drying treatments to produce the nanoparticles. This technique is great because it produces a very even distribution of nanoparticle sizes.

It's very difficult to take bulk materials and just grind them up to make nanoparticles; this method usually creates a broad particle size distribution of both nanoparticles and larger particles. One can isolate the nanosized particles by setting up a fluid flow in which the small particles get carried along and the larger ones fall out due to gravity, but it's a very tedious and slow process.

Why are gold nanoparticles used in so many medical applications?

Gold nanoparticles are used for three main reasons: first, it's very easy to make small, stable nanoparticles of gold; second, nanogold has been around for hundreds of years and has been used in medicine since the

1800s, so we have a lot of experience working with it; and third, gold is generally inert, which means that instead of reacting with it or dissolving it, the body will leave it alone to do what we want it to do.

How close are scientists to making nanorobots that go inside the body?
It depends on what you think of as nanorobots. There is a lot of hype about miniature versions of metal robots running around in the body or wreaking havoc on the world, but this is not the reality. But if you think of a "robot" as a functionalized nanoparticle that goes to a specific place in the body to do a specific thing, then nanorobots already exist; one example is drugs that are inserted into nanoparticles of clay because clay seems to go straight into the cells where the drug by itself wouldn't.

Could we feel nanorobots inside of us?
No, the particles don't have enough mass for us to feel them. Just as they are too small to see, they are too small to feel. In terms of what we would feel, they are not different from the other chemical substances we currently put into our bodies.

Are there any possible problems with using nanoparticles in sunscreens?
Although no adverse effects of nanoparticle sunscreens on humans have yet been found, there is concern that there may be undiscovered health issues for nanosized particles. The chemical industries have assumed that if large particles of a substance are safe, nanosized ones will be too, and the U.S. Food and Drug Administration (which must approve all active ingredients used in sunscreens sold the United States) bases its approval process on the identity and concentration of a chemical substance, not the size of the particle.

A growing body of evidence, however, suggests that the safety of nanoparticles can't be taken for granted just because larger particles of the same substance have been proven to be safe. Nanoparticles are useful precisely because they don't always act in the same ways as their larger counterparts. With sunscreens, the concern is that the nanoparticles could penetrate the protective layers of the skin and cause reactions with ultraviolet light that cause damage to DNA in cells. In 2003 the European Scientific Committee on Cosmetic and Non-Food Products (SCCNFP) concluded that titanium dioxide nanoparticles are safe for cosmetic use, but suggested the need for more tests on the safety of zinc oxide nanoparticles.

In response to this issue, many researchers are calling for full examination and safety testing of nanoparticles as if they were completely new chemicals, as well as clear identification of nanoparticles in ingredients lists of consumer products.

For more information about the *NanoSense* program, see www.nanosense.org

NANO ACTIVITY: SUNSCREENS AND SUNLIGHT ANIMATIONS, ADAPTED FROM SRI INTERNATIONAL'S NANOSENSE UNIT

Student Instructions & Worksheet

This animation worksheet is best used as an in-class activity with small groups so that students can discuss the different things they notice in the animations.

Important: These models are meant to provoke questions and stimulate a discussion about how the scattering mechanism works and how a real-world phenomenon can be represented with a model. As with all models, they are imperfect representations and are not meant to be shown to students simply as an example of "what happens."

Introduction

There are many factors that people take into account when choosing which sunscreen to use. Two of the most important factors that people consider are the ability to block UV light and how the sunscreen appears on the skin (due to the interaction with *visible* light). You are about to see three animations that are models of what happens when sunlight (UV and visible rays) shines on:

- Skin without any sunscreen
- Skin protected by 200 nm ZnO particle sunscreen
- Skin protected by 30 nm ZnO particle sunscreen

Viewing the Animations Online:

To view the animations, have your students navigate to the Clear Sunscreen Animation Web page at http://nanosense.org/activities/clearsunscreen/sunscreenanimation.html

Downloading the Animations:

If you have a slow Internet connection or want to have a copy of the animation on your computers for offline viewing, go to the Clear

Sunscreen Materials Web page at http://nanosense.org/activities/clearsunscreen/ and download the files "sunscreenanimation.html" and "sunscreenanimation.swf" to the same folder. To view the animation, simply open the file "sunscreenanimation.html" in your Web browser.

Questions

1. Select the UVA and UVB wavelengths of light with no sunscreen and click the play button.
   a. What happens to the skin when the UV light reaches it?

      The skin is damaged.

   b. How is the damage caused by the UVA rays different from the damage caused by the UVB rays? (You may want to play the animation with just UVA or UVB selected to answer this question.)

      In the animation, UVB light causes a burn on the skin's surface and UVA light causes the breakdown in skin fibers deeper in the skin that leads to premature aging.

   c. Based on what you know about the different energies of UVA and UVB light, why do you think this might happen?

      The UVB light causes more immediate damage to the first cells it encounters because it is high energy. The UVA light is lower in energy and can penetrate deeper into the skin before it does damage. Both UVB and UVA light also can lead to DNA mutations that cause cancer, which is not shown in the animation.

2. Now leave UVA and UVB light selected and try playing the animation first with the 30 nm ZnO sunscreen and then with the 200 nm ZnO sunscreen.
   a. What kind of sunscreen ingredient is shown in each animation?

      The 30 nm ZnO is a nanosized inorganic ingredient.
      The 200 nm ZnO sunscreen is a traditional inorganic ingredient.

   b. What happens to the UV light in the animation of 30 nm ZnO particle sunscreen?

      The UV light is completely blocked via absorption.

   c. What happens to the UV light in the animation of 200 nm ZnO particle sunscreen?

      The UV light is completely blocked via absorption.

   d. Is there any difference in how the UV light interacts with the 30 nm ZnO particles versus the 200 nm ZnO particles? Explain why this is so, based on your understanding of how the sunscreens work to block UV light.

There is no difference in how the 30 nm and 200 nm ZnO particles interact with the UV light. This is because absorption depends on the energy levels in the substance, which are primarily determined by the substance's chemical identity, not the size of the particle.

e. Is there any difference in how the two kinds of UV light interact with the sunscreens? Explain why this is so, based on your understanding of how the sunscreens work to block UV light.

Both UVA and UVB light are fully absorbed because ZnO absorbs strongly for all wavelengths less than ~380 nm.
Students may point out that wavelengths of 380–400 nm are UVA light that might not be absorbed. This is true and can be discussed with the final question, which addresses the limitations of using models.

3. Select the visible light option and play the animation for each of the sunscreen conditions. What happens to the visible light in each animation, and what does the observer see?

a. Skin without any sunscreen

The photons of light pass through the air to the skin. At the skin's surface, most of the blue-green (~400–550 nm) wavelengths of light are absorbed by pigments in the skin, while the red-orange-yellow (~550–700 nm) wavelengths of light are reflected and reach the observer's eye. The observer sees the surface of the skin. (Different skin colors are caused by different amounts and types of the skin pigment melanin).

b. Skin with 200 nm ZnO particles sunscreen

The photons of light pass through the air and are refracted (bent) as they enter the sunscreen. They are then scattered by the ZnO particles multiple times until they emerge from the sunscreen and are again refracted (bent). Since large particles of ZnO scatter all wavelengths of light equally, all of the different photon wavelengths reach the observer, who sees an opaque white surface. (Note that even though the animation shows the different colored photons reaching the observer at different times, in reality there are many photons of each color reaching the observer at the same time.)

c. Skin with 30 nm ZnO particle sunscreen

The photons of light pass through the air and are refracted (bent) as they enter the sunscreen. They pass through ZnO particles without being scattered, and at the skin's surface, most of the blue-green (~400–550 nm) wavelengths of light are absorbed by pigments in the skin, while the red-orange-yellow (~550–700 nm) wavelengths of light are reflected. They then pass through the sunscreen again, and are refracted (bent) when they pass to the air, before they reach the observer's eye. The observer sees the surface of the skin and we say that the sunscreen is "clear."

4. When we make a model (such as one of these animations), we make tradeoffs between depicting the phenomenon as accurately as possible and simplifying it to show the key principles involved. What are some other ways these animations have simplified the model of the real-world situation they describe?

Example Simplifications:

- The UVA and UVB light are each shown as two identical photons when in reality there are many more photons involved.
- The wavelengths of the two photons used to represent UVA and UVB light are shown to be the same when in reality each consists of a range of wavelengths.
- The ZnO particles are shown as "solid" balls when in reality they are clusters of ions.
- All of the ZnO particles are shown to be the same size, but in reality, there is a distribution of particle sizes.
- The damage of the UV rays to the skin doesn't show the DNA mutations that lead to cancer because of the size and timescale involved.
- The sunscreen solvent is a pale yellow, but it should be clear since it does not scatter (or absorb) light. How else could this be shown in the animations?

For the full version of this activity with all ten questions, please go to http://www.nanosense.org/activities/clearsunscreen/index.html and download the Teacher Materials for Lesson 4: How Sunscreens Appear: Scattering.

READING MATERIAL

Atkinson, William Illsey. *Nanocosm: Nanotechnology and the Big Changes Coming from the Inconceivably Small.* New York: AMACOM/American Management Association, 2003.

Drexler, Eric K. *Nanosystems: Molecular Machinery, Manufacturing, and Computation.* New York: John Wiley & Sons. 1992.

Fishbine, Glenn. *The Investor's Guide to Nanotechnology & Micromachines.* New York: John Wiley & Sons, 2001.

Hamakawa, Yoshihiro. *Thin-Film Solar Cells.* New York: Springer-Verlag, 2004.

Luryi, Serge, and Jimmy Xu. *Future Trends in Microelectronics: Reflections on the Road to Nanotechnology.* Boston: Kluwer Academic Publishers, 1996.

Scientific American (authors). *Key Technologies for the 21st Century: Scientific American: A Special Issue.* New York: W.H. Freeman & Co, 1996.

Uldrich, Jack, and Deb Newberry. *Next Big Thing Is Really Small: How Nanotechnology Will Change the Future of Your Business.* New York: Crown Publishing Group, 2003.

VIDEOS

Exploring the Nanoworld, movies of nano-structured materials including ferrofluids, memory metals, LEDs, self-assembly, and Lego models. National Science Foundation supported Materials Research Science and Engineering Center on Nanostructured Materials and Interfaces at the University of Wisconsin-Madison. http://mrsec.wisc.edu/edetc

G Living.The Phoenix Electric Nano Battery SUV. http://www.youtube.com/watch?v=w-Zv5RFgmWY&NR

Electron-Beam Lithography. Nanopolis Online Multimedia Library. Electron-beam lithography is a technique for creating extremely fine patterns required for modern electronic circuits. http://online.nanopolis.net/viewer.php?subject_id=139

Cosmetics. Nanopolis Online Multimedia Library. The cosmetics industry was one of the first industries to employ nanotechnology for cosmetics that include creams, moisturizers, and sunscreens. http://online.nanopolis.net/viewer.php?subject_id=274

WEB SITES

Nano Science and Technology Institute: http://www.nsti.org/

NanoBusiness Alliance: www.nanobusiness.org/

Center for Responsible Nanotechnology *Newsletter:*
http://responsiblenontechnology.org/newletter.htm

Nanomagazine (founded 2001, many interviews with leading figures): http://www.nanomagazine.com

Nano Technology Magazine: http://nanozine.com

Small Times: Daily articles covering MEMS, nanotechnology, and microsystems, with a business angle. http://www.smalltimes.com

NanoInk, Inc.: Creator of Dip Pen Nanolithography (DPN) tools for fabricating MEMS and other nanoscale devices. Chicago. http://www.nanoink.net

NanotechNews: News for nanotechnologists and investors: regular updates, many links to other nanotechnology Web sites, archives, search function, newswire. http://news.nanoapex.com/

SOMETHING TO DO

Self-cleaning glass is a nanotechnology consumer product that is available today. But, what is self-cleaning glass and how does it work? For an activity to help you learn more about the product, go to the NNIN Nanotechnology Web site: http://www.nnin.org/nnin_edu.html

7

Nanotechnology for Food, Agriculture, Livestock, Aquaculture, and Forestry

Nanotechnology has the potential to provide the tools and the research to change the future of food technology. Applying the principles of nanotechnology, researchers can produce more nutritious food and beverages; improve food packaging, and develop special biosensors. These biosensors can monitor food safety and the health of crops, forest areas, fish ponds, and livestock.

Did you know?

Due to food shortages, millions of people throughout the world suffer from malnutrition. In fact, malnutrition is a contributor to more than half the deaths of children under five in developing countries. One of the goals of the United States Department of Agriculture is to reduce hunger in America and the world.

According to one report, more than 200 companies worldwide are engaged in nanotech research and development related to food. The United States is the leader followed by Japan and China. All these countries and others will take part in a nanofood market that will surge from $2.6 billion today to $20.4 billion in 2010.

UNITED STATES DEPARTMENT OF AGRICULTURE

Nanotechnology is one of the U.S. government's top research priorities. The United States Department of Agriculture (USDA) is one of the leading agencies in supporting nanotechnology applications. The Department of Agriculture's multifaceted mission is to: ensure a safe food supply; care for agricultural lands, forests, and rangelands; support sound development of rural communities; provide economic opportunities for farm and rural residents; expand global markets for agricultural and forest products and services; and work to reduce hunger in America and the world.

The USDA investment in nanotechnology research has the potential to have a major impact on agriculture and food processing. One of their goals is to develop "smart" food packaging with built-in nanosensors to detect pathogens or contaminants in the product.

FOOD PACKAGING: A MAJOR GOAL USING NANOTECHNOLOGY METHODS

Today, food packaging and safety are the major goals of food-related nanotech research and development or R&D. Food companies are working on developing packaging that incorporates lighter and stronger packages with embedded sensors that can alert a consumer to contamination and or presence of pathogens.

According to industry analysts, the current U.S. market for the nanotechnology of packaging of foods and beverages is an estimated $38 billion, and will grow to $54 billion before 2010. The following includes a few examples that show nanoscale applications for food and beverage packaging.

Nano Plastic Packaging

Several chemical companies are producing a nanotechnology transparent plastic film for packaging containing nanoparticles of clay. The nanoparticles are integrated throughout the plastic and are able to block oxygen, carbon dioxide, and moisture from reaching fresh meats or other foods. The nanoclay particles also make the plastic lighter, stronger, and more heat-resistant.

One company that makes camera film used nanotech to develop antimicrobial packaging for food products. The company is also developing other "active packaging" that absorbs oxygen. If oxygen gets into packaged food products, such as cheese or meat, it can reduce the freshness of the food, so it may no longer be safe to eat.

Did you know?
Some food companies use oxygen absorbents to prevent the growth of aerobic pathogens and organisms that may spoil food or allow them to become unsafe.

Another manufacturer of plastics has developed a nano-composite plastic barrier for food that prevents the flow of oxygen and carbon dioxide from entering a package. The nanoplastic package keeps all kinds of products fresher longer. The plastics also block out smells. The plastic barrier is so strong that it can be useful in making boil-bag food products and microwavable packaging.

New Kinds of Bottles

Packaging companies are also interested in developing bottles that are as hard as glass but will not shatter if mishandled. The plastic bottles contain clay nanoparticles that are as hard as glass but far stronger, so the bottles are less likely to shatter. The layout of the nanoparticles is keeping the liquid from spoilage and flavor problems and giving it up to a six-month shelf life. By embedding nanocrystals in plastic, researchers have created a molecular barrier that helps prevent the escape of oxygen. Nanocor and Southern Clay Products are now working on plastic beverage bottles that may increase shelf life to 18 months.

Foodborne Diseases

The Centers for Disease Control and Prevention (CDC) estimates 76 million people suffer from foodborne illnesses each year in the United States, accounting for 325,000 hospitalizations and more than 5,000 deaths. Foodborne disease is extremely costly. Health experts estimate that the yearly cost of all foodborne diseases in this country is 5 to 6 billion dollars in medical expenses and lost productivity.

There are more than 250 known foodborne diseases. They can be caused by bacteria, viruses, or parasites. Some of the foodborne diseases include botulism, E. coli, salmonellosis, and listeria monocytogenes.

Listeria Monocytogenes in Food

Listeria monocytogenes has recently been recognized as an important public health problem in the United States. Listeriosis is a serious infection caused by eating food contaminated with the bacterium.

Listeria monocytogenes is found in soil and water. Vegetables can become contaminated from the soil or from manure used as fertilizer. The bacterium has been found in a variety of raw foods, such as uncooked meats and vegetables, as well as in processed foods that became contaminated even after processing, such as soft cheeses and cold cuts.

E. coli Handheld Sensor Detecting Bacteria with Electromechanical Cantilevers. Chemical engineers have developed a sensor that can almost instantly detect the presence of E. coli. Science Daily. http://www.sciencedaily.com/videos/2006-11-09/

Salmonella Ilnesses from Food

The Centers for Disease Control and Prevention has estimated that salmonella infections are responsible for nearly 1.5 million illnesses

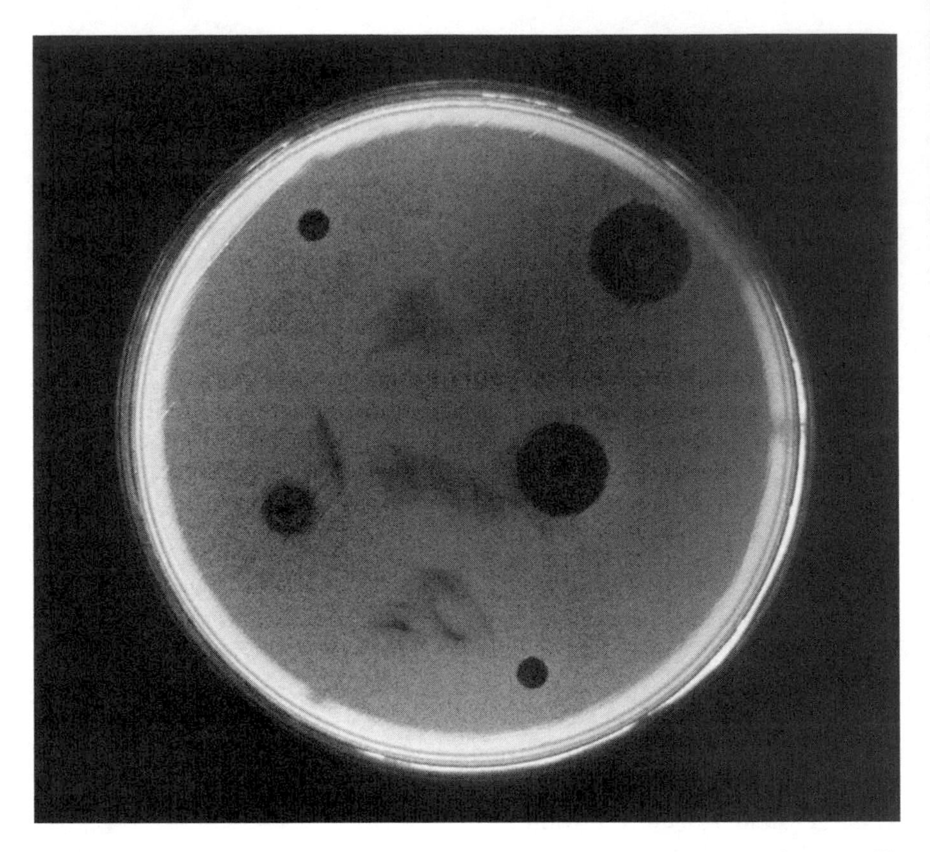

Escherichia coli is another foodborne pathogen that can cause human illnesses characterized by severe cramping (abdominal pain) and diarrhea. In this photo, *Escherichia coli,* being cultured in an agar plate, is the most common bacteria found in the large intestines of healthy individuals. (*Courtesy Centers for Disease Control*)

annually. *Salmonella* bacteria are often present in polluted water, on kitchen surfaces, in soil, within the bodies of insects or other animals, and on factory surfaces. Some of the most common vectors of *Salmonella* food poisoning are raw or undercooked meats, raw or undercooked poultry, eggs, milk and dairy products, fish, shrimp, yeast, coconut, and egg- or dairy-based sauces and salad dressings.

Infection with *Salmonella* bacteria is called salmonellosis. It is estimated that from 2 to 4 million cases of salmonellosis infection occur in the United States annually. This has resulted in more than 16,000 hospitalizations and nearly 600 deaths. The incidence of salmonellosis appears to be rising both in the United States and in other developed nations. Billions of dollars are lost in job productivity due to this disease.

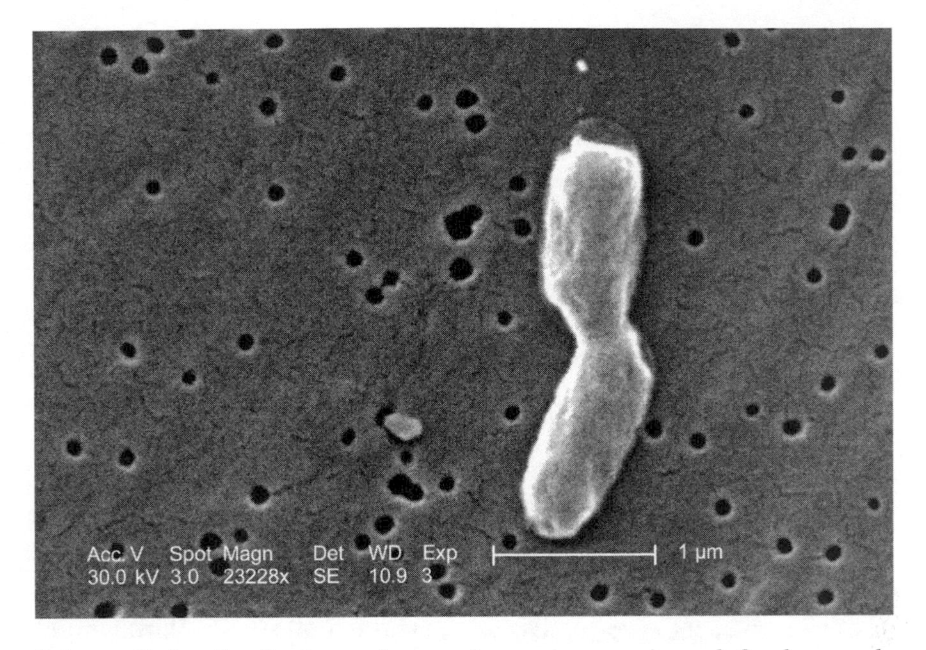

Salmonella is a foodborne pathogen that can appear in such foods as pork. The disease causes fever, abdominal cramps. The scanning electron micrograph shows two rod-shaped salmonella bacteria. (*Courtesy Centers for Disease Control*)

NANOSENSORS FOR FOODBORNE CONTAMINATION

Nanotechnology will provide safer and better quality foods for consumers by rapid detecting of pathogens in foods by using systems that use biosensors. By using biosensors, researchers and food processors can identify even tiny amounts of pathogens in food before products leave the food processing plants.

Professor Mohamed Ahmedna is a food scientist at North Carolina A&T State University. He and his research team are working on ways to detect contamination, such as salmonella, early so that the public can avoid foodborne illness.

Dr. Ahmedna describes the biosensor as, basically, a device that detects biological organisms such as salmonella or any other bacteria. The biosensors have the speed that you do not have with traditional methods. Speed is an issue as also are portability and cost.

Did you know?
Today sensors provide information about temperature and weather data, chemical contaminants, and even control the deceleration for release of airbags in automobiles.

This instrumentation may look simple, but tests to date have shown that it can rapidly detect salmonella. Currently, industry methods for food-borne bacteria detection can take several days. This biosensor unit, once refined, may deliver results in just a few hours.

USING FOOD PACKAGING SENSORS IN DEFENSE AND SECURITY APPLICATIONS

In the view of the U.S. military, it is a national security priority to detect foodborne pathogens. With the present technologies, testing for microbial food contamination takes 2 to 7 days for the results. The sensors that have been developed to date are too big to be transported easily. In the immediate future, small, portable sensors that can detect and measure pathogens in food products are needed.

Several groups of researchers in the United States are developing biosensors that can detect pathogens quickly and easily. These sensors can play a critical role in the event of a terrorist attack on the food supply. With United States Department of Agriculture (USDA) and National Science Foundation (NSF) funding, researchers at Purdue University are developing a handheld sensor capable of detecting a specific bacteria instantaneously from any sample.

Detecting Deadly Chemicals. A sampler gun that can now be used to detect harmful or dangerous diseases essuch as anthrax. ScienceDaily. http://www.sciencedaily.com/videos/2006-12-10/

OTHER KINDS OF SENSORS: THE ELECTRONIC NOSE AND THE ELECTRONIC TONGUE

Two sensors, called the electronic nose and the electronic tongue, can detect food pathogens. Both these sensors have been used in laboratories, including NASA, to detect trace chemicals. Now, the electronic nose and the electronic tongue will be applied to the food industry, since both sensors are especially effective in recognizing contaminants and analyzing the overall quality of foods.

The Electronic Nose (e-nose) Technology

An electronic nose (e-nose) is a device that identifies the specific components of an odor and analyzes its chemical makeup to identify it. In a sense it mimics the human nose, but it has far greater sensitivity to smell and can trace scents at the nanoparticle level.

Electronic noses are not new. They have been around for several years. Electronic noses were originally used for quality control applications in the food, beverage, and cosmetics industries. However, the e-nose is large and expensive to manufacture.

Current research is focused on making the e-nose smaller, less expensive, and more sensitive. The new and improved e-nose could be used to detect odors specific to diseases for medical diagnosis, the detection of pollutants, and gas leaks for environmental protection.

The Electronic Tongue Technology

Similar to the e-nose technology, the electronic tongue, or e-tongue, mimics the human tongue, but it is more sensitive to flavors in foods. The e-tongue sensors can detect substances in parts per trillion and could be used in packaging to trigger a color change that would alert the consumer if a food had become contaminated or if it had begun to spoil.

Food researchers believe the electronic tongue is going to be vital to food studies. As an example, a meat package with a built-in tongue can taste the first signs of spoilage and activate a color change as a warning to the consumer.

The electronic tongue is made from a silicon chip that has microbeads arrayed on it. Each of the beads responds to different analytes. The different analytes of the e-tongue are similar to the taste buds on the human tongue. The e-tongue responds to sweet, sour, salty, and bitter tastes in a similar fashion as the taste buds on your tongue do.

The food and beverage industries may see the potential to use the e-tongue to develop a digital library of tastes. The collected data would include tastes that have been proven to be popular with consumers. The e-tongue could also monitor the flavors of existing products. Probably more importantly, the e-tongue could be used for brand security and tracking the supply chain of foods that normally cannot be tagged with traditional bar codes.

NANO BAR CODES DETECT FOODBORNE DISEASES

Another kind of sensor technology is the nano bar code. Nano bar codes are similar to the traditional bar codes that are found on many packaged food products today. However, the nano bar codes, containing metal nanoparticles, could be used to detect pathogens. The nanoparticles have specific, recognizable chemical fingerprints that can be read by a machine, an ultraviolet lamp, or an optical microscope. Using these kinds of bar codes, a supermarket checkout computer can identify thousands of different items by scanning the tiny bar code printed on the package. The scanning of the barcode could reveal spoilage or pathogens within the food package.

A research group headed by Dan Luo, Cornell assistant professor of biological engineering, has created "nanobarcodes." The nanobarcodes fluoresce under ultraviolet light in a combination of colors that can be read by a computer scanner or observed with a fluorescent light microscope.

Summarizing, nanosensors such as electronic tongues and electronic noses plus the use of nanobarcodes have the potential to provide safe food products and packaging. The embedded sensors in food packaging will respond to the release of particular chemicals when a certain food begins to spoil. So, as soon as the food starts to go bad, the packaging will change color to warn the shopper. This system could also provide a more accurate and safer method than the present "sell by dates" marked on food products.

AGRICULTURE AND NANOTECHNOLOGY

The application of pesticides has become a controversial topic due to claims that some products can damage the environment and even get into the food chain. As an example, it has been estimated that about 2.5 million tons of pesticides are used on crops each year. The worldwide damage caused by pesticides reaches $100 billion annually. One of the reasons for the damage is the high toxicity and nonbiodegradability of pesticides. Pesticides that are applied directly to the soil may be washed off the land into nearby bodies of surface water or may percolate through the soil to lower soil layers and groundwater. When this happens, the pesticide becomes a pollution hazard and is highly toxic to humans and other organisms.

BIOSENSOR DETECTS HERBICIDES ON THE FARM

Researchers are hoping a new biosensor may help farmers and regulatory officials detect herbicides in soil and water samples. Herbicides are used to control unwanted plants such as weeds. However, heavy applications of herbicides can leave environmentally unsafe residues in soil and water.

The new biosensor is made of a chlorophyll-like chemical that can measure oxygen levels. This chemical produces oxygen in the presence of certain chemicals and light.

In testing for a herbicide, a liquid sample is passed through the biosensor. If the sample contains a herbicide, it will react with the biosensor's

Table 7.1 Pesticide Persistence in Soil

Low (half-life < 30 days)	Moderate (half-life 30–100 days)	High (half-life > 100 days)
Aldicarb	Aldrin	Bromacil
Captan	Atrazine	Chlordane
Calapon	Carbaryl	Lindane
Dicamba	Carbofuran	Paraquat
Malathion	Diazinon	Picloram
Methyl-parathion	Endrin	TCA
Oxamyl	Heptachlor	Trifluralin

proteins and inhibit oxygen production. The electrode in the biosensor detects oxygen levels and sends the information to a computer that displays the data in graph form. Reading the oxygen levels from the data a scientist can determine if too much or too little of herbicides has been applied to the soil. The biosensor can also identify traces of chemical residues in a matter of minutes.

Nanoscale Herbicides

Other researchers are focusing on ways to reduce the use of herbicides all together. Their research includes using nanoparticles to attack the seed coating of weeds, which will prevent them from germinating.

The researchers report that this approach will destroy the weed even when it is buried in soil, and will prevent them from growing even under the most favorable conditions. They believe this method is more preferable to tilling and manual picking of weeds because of the costs incurred with such high-maintenance methods. Using small proportions of nanoscale herbicides, the nanoparticles can easily blend with soil and attack weeds that are buried below the reach of tillers and conventional herbicides. More research has to be conducted to make sure the nanoparticles can be used safely in the soil.

> **Did you know?**
> Two other major groups of pesticides, besides herbicides, include insecticides that are used to control insect damage and fungicides that are applied to kill fungi and other parasites.

A FOOD SAFETY ISSUE

The issue of food safety has also been heightened by the possibilities of bioterrorism. Global businesses, government agencies, and other organizations are now looking at a variety of technologies, as well as nanotechnologies that can trace food distribution from the farms, the food-processing companies, to the stores and supermarkets.

ATOMIC FORCE MICROSCOPY AND FOOD RESEARCH

An estimated 25–81 million cases of foodborne illness and an estimated 9,000 deaths are associated with consumption of contaminated foods each year. Therefore, the presence of microbial and chemical pathogens at any stage of food production, processing, and distribution must be quickly determined in order to allow proper treatments before food consumption by the general public. To help achieve these goals, the United States Department of Agriculture (USDA) and other agencies use the Atomic Force Microscope (AFM). The AFM is implemented in food science research to observe the nanoscale structure of foods and other biomaterials and for investigating diseases, such as E. coli cells and salmonella.

SUSTAINABLE WATERING OF CROPS

Agricultural scientists are exploring different ways to use water in a more sustainable way for irrigation. Presently, the main disadvantage of using irrigation is that it requires a lot of water, much of which is lost to evaporation. In some farm areas, more than 50 percent of water may be lost to evaporation. The high evaporation rate may also decrease soil fertility through salinization.

Nanotechnology researchers would like to come up with a way to apply water slowly into the ground with a minimum of runoff and evaporation. One idea is to use the mineral zeolites to do the job.

Zeolites are microporous crystalline minerals that contain silicon, aluminium, and oxygen. Zeolites are used in petroleum refineries and for water softening and purification products.

The researchers are interested in using zeolites in water irrigation systems because the mineral can absorb about half its own volume of water. The zeolites would act as water moderators whereby they will absorb up to 50 percent of their weight in water and slowly release it to plants.

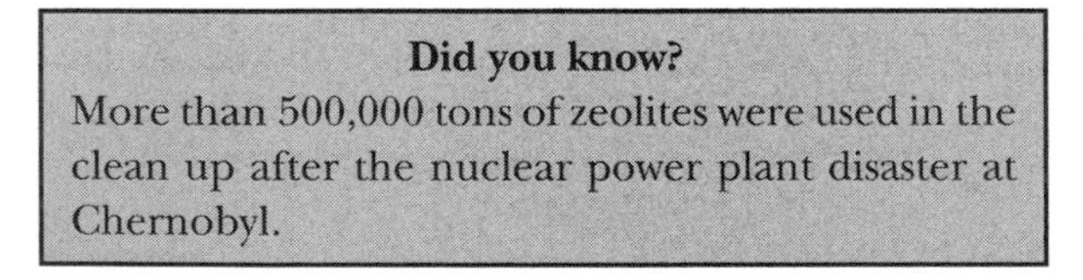
Did you know?
More than 500,000 tons of zeolites were used in the clean up after the nuclear power plant disaster at Chernobyl.

Using zeolites in a water irrigation system has several advantages over other conventional irrigation methods. First, the zeolites could reduce the total amount of water needed to irrigate crops, making this kind of irrigation method useful in regions with limited water supplies. Second, little water is lost to the air through evaporation or to the soil via percolation because water is released slowly and over a small area. At the same time, fewer salts are deposited in the soil, helping to preserve soil quality.

LIVESTOCK DISEASES

The worldwide livestock industry is interested in measures that would guarantee the safety of the food supply provided by livestock. Outbreaks of livestock disease have resulted in export bans and collapsed markets. As one example, in the United Kingdom, the Mad Cow Disease crisis in the late 1990s led to a 40 per cent domestic decline in beef sales and the complete loss of many export markets.

Mad cow disease is the commonly used name for Bovine Spongiform Encephalopathy (BSE). The disease is a slowly progressive, degenerative, fatal disease affecting the central nervous system of adult cattle. The exact cause of BSE is not known.

Scientists are now exploring nanotechnology applications such as the use of small drug delivery devices. The devices could be implanted into cattle to avoid diseases such as BSE from entering the animals' central nervous system.

BIOCHIPS FOR DISEASE DETECTION IN LIVESTOCK

Scientists at the Kopelman Laboratory at the University of Michigan are developing noninvasive nanosensors to test livestock diseases. One of these nanosensors could perhaps be placed in, say, a cow's saliva gland in order to detect single virus particles long before the virus has had a chance to multiply and long before disease symptoms are evident.

Researchers at the University of Pretoria, for example, are developing biochips that will detect common diseases borne by ticks. A biochip (or microarray) is a device typically made of hundreds or thousands of short strands of artificial DNA deposited precisely on a silicon circuit. Chips can be used for early disease detection in animals.

Did you know?
Avian influenza, or "bird flu," is a contagious disease of animals caused by viruses that normally infect only birds and, less commonly, pigs.

Biochips can also be used to trace the source of food and feeds. For example, a biochip tests livestock feed to detect the presence of animal products from forty different species as a means to locate the source of pathogens—a response to public health threats such as avian flu and mad cow disease.

Biochips for Animal Breeding

One goal is to functionalize biochips for breeding purposes. With the mapping of the human genome behind them, geneticists are now rapidly sequencing the genomes of cattle, sheep, poultry, pig, and other livestock hoping to identify gene sequences that relate to commercially valuable traits, such as disease resistance, and leanness of meat. By including probes for these traits on biochips, breeders will be able to identify champion breeders and screen out genetic diseases.

NANOSENSORS TO TRACK LIVESTOCK

Livestock tracking has been a problem for farmers. The USDA envisions the rise of "smartherds"—cows, sheep, and pigs fitted with sensors and locators relaying data about their health and geographical location to a central computer.

Implanting tracking devices in animals has been initiated for pets, valuable farm animals, or for animal wildlife conservation. Injectable microchips are already used in a variety of ways with the aim of improving animal welfare and safety to

- study animal behavior in the wild,
- track meat products back to their source, and
- reunite strays with their human owners.

NANOTECHNOLOGY IN AQUACULTURE AND FISH FARMING

Aquaculture is the cultivation of fish, shellfish, or aquatic plants in natural or controlled marine or freshwater environments. Estimates report that about 20 percent of all commercial fish are raised in an aquaculture environment, and that this industry will continue to grow in the 21st century.

Today, aquaculture is a multimillion dollar business. Much of the trout, catfish, and shellfish consumed in the United States are products of aquaculture. Global demand for seafood is projected to increase by 70 percent in the next 30 years. In fact, the world's fastest growing area of animal production is the farming of fish, crustaceans, and mollusks, particularly in Asia. A consensus is growing that a dramatic increase in aquaculture is needed to supply future aquatic food needs.

Environmental Concern of Aquaculture

However, there are problems with some of the fish farms, particularly shrimp-raising farms. Disease outbreaks, chemical pollution, and the environmental destruction of marshes and mangroves have been linked with fish farm activities. Environmentalists believe more sustainable practices are needed to control the potential for pollution and the damage of natural resources. There is also concern about the accidental releasing of cultivated fish into natural populations and what effects that might have.

Cleaning Fish Ponds with Nanotechnology Devices

Researchers believe that nanotechnology may have the potential to provide fishponds that are safe from disease and pollution. One company, Altair Nanotechnologies, makes a water-cleaning product for fishponds called *NanoCheck*. It uses 40-nanometer particles of a lanthanum-based compound that absorbs phosphates from the water and prevents algae growth.

Altair sees a potentially large demand for *NanoCheck* for use in thousands of commercial fish farms worldwide where algae removal and prevention is costly at present. According to Altair, the company plans to expand its tests to confirm that its nanoparticles will not harm fish.

DNA Nanovaccines Using Nanocapsules and Ultrasound Methods

The USDA is completing trials on a system for mass vaccination of fish in fishponds using ultrasound. Nanocapsules containing short strands of DNA are added to a fishpond where they are absorbed into the cells of the fish. Ultrasound is then used to rupture the capsules, releasing the DNA and eliciting an immune response from the fish. This technology has so far been tested on rainbow trout by Clear Springs Foods (Idaho, US)—a major aquaculture company that produces about one-third of all U.S. farmed trout.

FOREST PRODUCT INDUSTRY AND NANOTECHNOLOGY

According to recent research, the future of the U.S. forest products industries, which employ some 1.1 million Americans and contribute more than $240 billion annually to the nation's economy, could depend on how well those industries embrace the emerging science of nanotechnology.

The hundred-page report, titled "Nanotechnology for the Forest Products Industry—Vision and Technology Roadmap," is based on presentations and discussion by some 110 researchers from North America and Europe, with an interest in wood, paper, or other forest products, to explore the possible role of nanotechnology in the forest products industries.

The "Roadmap," so called because it is intended to show where the forest products industry needs to go and how to get there, describes the U.S. forest products industry as a mature, somewhat stagnant energy-intensive industry that is facing new global competition. The report, which is the first comprehensive look at nanotechnology for the U.S. forest products industry, suggests that the infusion of nanotechnology could lead to new and improved products. It also could lead to improved and more efficient manufacturing processes.

"Nanotechnology represents a major opportunity to generate new products and industries in the coming decades," the Roadmap says.

Potential uses of nanotechnology in forest products, as identified in the Roadmap, include development of intelligent wood- and paper-based products that could incorporate built-in nanosensors to measure forces, loads, moisture levels, temperatures, or pressures, or detect the presence of wood-decay fungi or termites. According to the Roadmap, nanotechnology can have an even greater impact by providing benefits that extend well beyond food products but into the areas of sustainable energy production, storage, and utilization.

ENSURING NANOTECHNOLOGIES IN FOOD PRODUCTION TO BE SAFE

Many groups are concerned about the potential effects on human health and the environment of manufactured nanoparticles in foods and food preparation. To establish appropriate safety tests and clear labeling, more research is needed to ensure that nanotechnologies are developed in a safe and socially acceptable way.

> **Did you know?**
> Synthetic zeolites are widely used as catalysts in the petrochemical industry.

NANO INTERVIEW: PROFESSOR MAKARAND (MAK) PARANJAPE, Ph.D., GEORGETOWN UNIVERSITY

As you learned in earlier chapters and in this one, biosensors have the potential to be used in several fields including human health and food preparation. In this interview, Dr. Paranjape discusses his work

Professor Paranjape's Micro and NanoSystems Group. From left to right: Jonathan Hesson (researcher), Jianyun Zhou (Ph.D. candidate), Vincent Spinella-Mamo (Ph.D. candidate), Yogesh Kashte (researcher), Sean Flynn (undergraduate sophomore student), Mak Paranjape (group leader). Missing from picture: Andrew Monica (Ph.D. candidate), Megan Giger (undergraduate junior student) and Rajeev Samtani (high school senior). (*Courtesy Russel Ross*)

in developing a biosensor that detects glucose levels in patients with diabetes.

Dr. Paranjape is an Associate Professor in the Department of Physics at Georgetown University (Washington, DC), where he joined the faculty in 1998. He received his Ph.D. in Electrical Engineering from the University of Alberta (Edmonton) in 1993, and held postdoctoral researcher positions at Concordia University (Montreal), Simon Fraser University (Vancouver), and the University of California (Berkeley). In 1995, through a collaborative project with U.C. Berkeley, Dr. Paranjape joined the Istituto per la Ricerca Scientifica e Tecnologica (IRST) in Trento, Italy, as a research consultant.

Dr. Paranjape and a team at Georgetown University and Science Applications International Corporation (SAIC) have developed a biosensor micro-device that has the potential to be used by people who have diabetes. The biosensor is in the form of a small adhesive patch to be worn on the skin and is very convenient to use and makes measuring glucose levels completely pain-free and in a minimally invasive manner. The multidisciplinary project was funded for over 3 years with technical expertise coming from several scientific backgrounds: one other physics professor, one professor in pharmacology, two senior researchers from SAIC (biochemistry and engineering), six postdoctoral researchers (chemistry, materials engineering, biochemistry, microfabrication specialists, and two electrical engineers), and several undergraduate and graduate students in the physics program.

Where did you grow up and what were some of your favorite activities and subjects as a teenager?
I was born in England but I grew up in Canada and lived as a teenager in a small town called Thunder Bay, which is right on Lake Superior, almost 30 miles from the United States border north of Minnesota. I enjoyed racquet sports, skiing, and playing hockey—my favorite position was playing goal.

What colleges did you attend and what was your major?
I completed my undergraduate and graduate degrees, including my Ph.D., all in electrical engineering, at the University of Alberta in Edmonton, Canada. It was during this time when Wayne Gretzky, one of my hockey idols, was playing for the Edmonton Oilers, a National Hockey League team that won several back-to-back Stanley Cups while I was in school.

Presently, what subjects do you teach at Georgetown University?
I have an electrical engineering degree, so I teach related subjects such as electricity and magnetism, and electronics to both physics undergraduates and graduate students.

What interested you to become a physics professor, and how did you cross over and get interested in the field of diabetes and in being on a team that developed a new biosensor device for glucose monitoring?
My early interests were to follow in my dad's footsteps. He is a physicist by training. I also thought about medicine, as well. I liked math a lot and enjoyed hands-on work, but my high school biology skills were not that great. So I wanted to go into physics, but my dad said there are not that many jobs (or money) in this field. He thought that since I was math-oriented and a hands-on type person, I should consider engineering. So that is how I ended up in electrical engineering.

How did you get involved in biosensors?
The last time I took a biology course was in the 12th grade and I have not taken any biology courses since then.

After I finished my Ph. D., I started to dabble in biology and biosensors during a time when I was employed as a postdoctoral researcher at Simon Fraser University. During this time, while I was involved in a project to accurately measure the mass of individual cells, such as baby hamster kidney cells, using a relatively new emerging technology called biological microelectromechanical systems (or bio-MEMS), I became very interested in fabricating devices using microtechnologies, whether they are sensors or "machines" that could interact with biological cells. This work was very intriguing to me. Anyway, this is how I got into biosensors. I got involved with the glucose-monitoring project soon after arriving at Georgetown University in 1998. Our research team was already involved in fabricating noninvasive microdevices for health applications, and after winning a U.S. Defense Department contract for monitoring glucose and lactate in a soldier, the minimally-invasive glucose patch was developed.

What is the conventional way a person living with diabetes uses to measure and monitor his or her glucose level? Approximately, how many times a day does a person have to do this routine?
Typically a person with diabetes, and depending on the severity of the disease, should sample his or her glucose level at least twice a day. And that is the minimum. Typically it is done 3 to 5 times a day—once when you get up, once after breakfast, once after lunch, and once after dinner, and finally once before bedtime. The more often you monitor your readings, the better control you have on the disease. However, the more you monitor, the more painful the process can be for many people with diabetes. Remember you are sticking a needle in your finger to draw blood and there is always going to be pain associated with this approach. And one of the big issues related to diabetes control is that many people, because of the pain, will not sample themselves the number of times they should in order to control their diabetes. Then after getting a blood sample, and depending on the glucose reading, a person would inject insulin, if needed. And this can occur after each glucose reading.

How does the new biosensor device work without puncturing the skin and what are the benefits of the device for patients?
For the past few years, the team has been developing and testing a new biosensor device for glucose monitoring. The size of a small band-aid, it is designed to be worn anywhere on the body, where the biosensor

samples tiny amounts of fluids that lie just beneath the top layer of skin.

The device is small and convenient, and makes measuring glucose levels pain-free and minimally invasive.

The biosensor device works to painlessly remove the outer dermis, or dead-skin layer, by using a "micro-hotplate." The hotplate temperature is carefully controlled to apply a small amount of power. The "hotplate" is turned on to a temperature of 130°C. This sounds hot, but in such a small spot the size of a hair follicle, and for such a short time, a person cannot even detect the heat, or feel any pain, as it is applied to the very outer layers of skin. The biosensor then determines the glucose levels from the sample of fluid, which rises to the skin surface through the micro-pores created by the hotplate, using tiny micro-electrodes that have been coated with a substance that reacts specifically to the glucose. The fluid being sampled is interstitial fluid and not blood, which is commonly used to monitor glucose levels.

Can young people who are 12 years old or younger use the biosensor device? Are there any age limits?

Any age group can use the biosensor. The ultimate goal of the device is not just to monitor the glucose level but eventually, to take steps to actually inject insulin. The biosensor will form a complete closed loop system so the device will effectively monitor glucose levels and deliver any insulin if needed. In a sense the envisioned device would perform as both biosensor and drug delivery system for the person with diabetes.

How long does the present biosensor patch last on the body?

The present patch contains several sensing sites and each of these sites provides a one-time glucose reading. When you are out of sensing sites, then the patch is removed. And this depends on how many times you are sensing your glucose level. The patch, when marketed, could last for about two weeks depending on the person who will test their glucose levels say 2 to 3 times a day. However, if you were able to add more and more sensing sites, then a patch could last much longer.

Is the biosensor device still being tested, and if so, when do you think the device would be available people with diabetes?

Usually there is a long process needed for FDA approval of any new drug or new medical devices. However, the patch has some advantages in that it is a minimally invasive patch because you do not need to implant it and you do not need to have to collect blood. Typically these kinds of devices are fast-tracked through FDA approval. Once the devices have been fully tested using animals, and then after performing clinical trials on human subjects, you can apply for FDA approval. It will take about 2 years more for testing on humans. And then the fast track will take about 1 to 2 years for FDA approval.

What are some other possible applications for the biosensor device?
Let me start by saying this project was sponsored by the Department of Defense, DARPA (Defense Advanced Research Projects Agency). They were interested in using a patch technology, similar to our designed biosensor patch. The patch would be for the war fighter on the battlefield who could succumb to trauma due to huge losses of blood. Blood lost is critical so you need to minimize the blood lost as quickly as you can to have a chance of saving the injured trooper. In this application, troops would be sent onto a battlefield, fitted with a biosensor patch placed on their arms. The patch would not hinder them from continuing their duties.

If injured, the patch would record data about the severity of the trooper's wounds based on their glucose and lactate levels (and any other readily available bio-molecule that can be sampled with our patch). The data would be sent by short-range telemetry to a medic's palm pilot, for example. The palm pilot would show the location of the injured trooper. The severity of the injury would also be ascertained. If the medic observed dangerous changes in a soldier's body biochemistry, that soldier would be assisted first, and then removed from the field for medical care. So in our continued research, we incorporated the military application with the civilian application kept in mind—that is, detecting glucose in people with diabetes.

Do you see other uses of this kind of biosensor?
There is a whole list of applications. One other area of interest is to detect, in newborns, the potential of jaundice before its onset. The new mother and baby would leave the hospital and a biosensor patch would be placed on the child's arm. If any signs of jaundice would appear when the child is at home, the patch, which monitors a biomolecule called bilirubin, would signal an alert. In this way, you would catch the jaundice earlier and then proceed with medical care. The patch would perform the sensing without pain, unlike current methods that involve inserting a heel stick to draw blood from the newborn's heel.

What advice would you give young people who would like a career in biomedical research developing biosensors and other microdevices for diagnostics?

In high school, there is not much overlapping of subjects between the various science disciplines. However, at the college level, and in the real world, there is a lot of interaction and collaboration that takes place between people who have various backgrounds in physics, chemistry, biology, mathematics, and engineering. You do not need to be an expert in everything. You can have your own expertise in one subject but always consult others in different fields to see what else can be done.

You can watch a short film of Prof. Paranjape and the GAEL lab where the biosensor device is produced. The film is called, *Monitoring Blood Glucose Without Pain or Blood.* The Web site is: http://college.georgetown.edu/research/molecules/14887.html

NANO ACTIVITY: FOOD PACKAGING

Since nanotechnology will be applied to the manufacturing of food packages you may be interested in learning more about food packaging by doing activities from The NSF-funded Materials World Modules MWM) Program. One of the interdisciplinary modules is called *Food Packaging.* In the *Food Packaging* module, you can learn about the many functions of food packaging and how these materials affect the environment. You can design your own environmentally friendly package for delivering a hot baked potato and its topping.

See the video of one teacher's experience in using the modules at http://www.materialsworldmodules.org/videocenter/MWMinterview_kh.htm

For more information, contact: Materials World Modules, Northwestern University, 2220 Campus Drive, Cook Hall, Room 2078, Evanston, IL 60208 or mwm@northwestern.edu

READING MATERIALS

Goodsell, D.S. *BioNanotechnology: Lessons from Nature.* Hoboken, NJ: Wiley-Liss, Inc., 2004.

Gross, Michael. *Travels to the Nanoworld: Miniature Machinery in Nature and Technology.* New York: Perseus Books Group, 2001.

Mulhall, Douglas. *Our Molecular Future: How Nanotechnology, Robotics, Genetics, and Artificial Intelligence will Transform Our World.* Amherst, NY: Prometheus Books, 2002.

National Research Council. *Frontiers in Agricultural Research: Food, Health, Environment and Communities,* Committee on Opportunities in Agriculture, Washington, DC: National Academies Press, 2003.

Schnoor, J.L. (Editor). *Fate of Pesticides and Chemicals in the Environment* (Environmental Science and Technology), New York: John Wiley & Sons, 1991.

The United States Department of Agriculture. National Planning Workshop. *Nanoscale Science and Engineering for Agriculture and Food Systems.* Washington, DC: 2003.

VIDEOS

Monitoring Blood Glucose without Pain or Blood. Go to: http://college.georgetown.edu/research/molecules/14887.html

Detecting Deadly Chemicals. Science Daily. Investigators on a crime scene can now use a new tool for collecting chemical or biological samples. The sampler gun collects samples on a cotton pad—eliminating direct contact with anything harmful, as well as risk of contaminating evidence—a GPS system to record the samples' location, a camera that snaps pictures for evidence, and a digital voice recorder and writing pad for taking notes. http://www.sciencedaily.com/videos/2006-12-10/

Detecting Toxics. A new portable lab that detects deadly chemicals in the air. ScienceDaily. http://www.sciencedaily.com/videos/2006-02-09/

Materials World Modules. Northwestern University. The NSF-funded Materials World Modules MWM) Program has produced a series of interdisciplinary modules based on topics in materials science, including Composites, Ceramics, Concrete, Biosensors, Biodegradable Materials, Smart Sensors, Polymers, Food Packaging, and Sports Materials. The following video describes the experiences of one teacher who used the Materials World Modules. Inside MWM: An Interview with Lake Forest Science Teacher Kate Heroux. In this interview with science teacher Kate Heroux, Professor Robert Chang, the developer of MWM, discusses the successes and challenges of implementing the Materials World Modules in high school. To see the video go to: http://www.materialsworldmodules.org/videocenter/MWMinterview_kh.htm

WEB SITES

Task Force on Building a Science Roadmap for Agriculture, National Association of State Universities and Land-Grant Colleges (NASULGC), Experiment Station Committee on Organization and Policy (ESCOP), "A Science Roadmap for the Future." November 2001. www.nasulgc.org/comm_food.htm

Nanotechnology for the Forest Products Industry—Vision and Technology Roadmap. www.fpl.fs.fed.us/highlighted-

National Science and Technology Council. *The National Nanotechnology Initiative: Leading to the Next Industrial Revolution,* Committee on Technology (Interagency Working Group on Nanoscience, Engineering and Technology, Washington, DC: 2000. www.ostp.gov/NSTC/html/iwgn/iwgn.fy01budsuppl/toc.htm

United States Department of Agriculture: http://www.usda.gov

U.S. Forest Service: http://www.fs.fed.us

SOMETHING TO DO

Soil Uses

Each state in the United States has selected a state soil, 20 of which have been legislatively established. Soils, because of their chemical and physical properties determine land use purposes, such as farming, mining, and ranching and even housing. For a science report, you may want to research and write about your state's soil and how the land is used for agriculture and other uses. You can write about topics that include soil quality, soil risks and hazards, and soil salinity and plant tolerance and how nanotechnology can impact these soil issues. To get information about your state soil go to: http://soils.usda.gov/gallery/state_soils/

8

Nanotechnology for a Sustainable Environment

We have learned from the previous chapters that applications of nanotechnology have the potential to change and improve everything from medical care and food safety to consumer products and electronics. In addition to these benefits, nanotechnology may help build a more sustainable environment by providing clean drinking water, better air quality, new sources of energies, and a reduction in hazardous and toxic wastes.

WATER POLLUTION AND NANOTECHNOLOGY

By 2015, approximately 3 billion people will live in countries where it will be difficult to get enough water for basic needs. More than 1 billion people will lack access to clean drinking water, while others will die from contaminated water. The Environmental Protection Agency (EPA) estimates that at least 500,000 cases of illnesses annually can be attributed to contaminated drinking water in the United States.

> Safer Water Worldwide. Industrial toxicologists developed a water purifier which separates particles and organisms from water and can be used to detect pesticides in water. Go to Web site. http://www.sciencedaily.com/videos/2006-12-06/

Water can be polluted by excessive amounts of chemicals such as: heavy metals, organic chemicals, disease-causing microorganisms, sediments, heat or thermal pollution and others. These sources of pollutants come from homes, mining activities, municipal sewage plants, industry and manufacturing plants, ranching, and agricultural operations.

> **Did you know?**
> Heavy metals include metallic elements such as mercury, lead, and arsenic that are hazardous and toxic and are a health risk for humans and other living organisms.

NANOTECHNOLOGY AND SAFER DRINKING WATER

Arsenic contamination, both naturally occurring and resulting from human activities, in drinking water is a global problem. For one example, in India, millions of people suffer from arsenic poisoning found in drinking water.

Cleaning Up Arsenic

Scientists are now developing strategies to clean up the arsenic, a carcinogen, in the water. Scientists at Rice University's Center for Biological and Environmental Nanotechnology (CBEN) have developed a low-cost technology for cleaning arsenic from drinking water. The technology holds promise for millions of people not only in India but also in Bangladesh and in other developing countries where thousands of cases of arsenic poisoning are linked to poisoned wells each year.

The scientists discovered that nanoparticles of iron oxide (rust) could be used to remove arsenic in water by using a magnetic field. Arsenic adheres to rust, according to the scientists. Rust is essentially iron oxide, a combination of iron and oxygen, and tends to be magnetic. The arsenic particles that stick onto the iron oxide can be removed from the water by using a low-powered magnet that attracts the particles. Once the particles are extracted, the water becomes safe to drink.

> **Did you know?**
> The following analogy will illustrate how little of Earth's freshwater is available: If all of Earth's water could fill a 4 liter (about 1 gallon) container, the amount of available fresh water would fill less than a tablespoon out of the 4 liters. As you can see, only a small percentage of freshwater on Earth is available for human use.

The Element Arsenic

The element arsenic is a steel grey metal-like material. Arsenic is a natural part of our environment and widely distributed in Earth's crust. So living organisms are often exposed to some amount of it. Very low levels of it are always present in soil, water, food, and air. Most arsenic compounds have no smell or special taste, even when present in drinking water. Arsenic has been used in pesticides, poisons, chicken-feed supplements, and wood preservatives. However, the naturally occurring arsenic in the soil can make the water toxic to humans.

Table 8.1 Water Pollutants

Point Sources	Bacteria	Nutrients	Total Dissolved Ammonia	Solids	Acids	Toxics
Municipal sewage treatment plants	•	•	•			•
Industrial facilities				•		•
Combined sewer overflows	•	•	•			•
Nonpoint Sources						
Agricultural runoff	•	•		•		•
Urban runoff	•	•		•		•
Construction runoff		•				•
Mining runoff				•	•	•
Septic systems	•	•				•
Landfill spills						•
Forestry runoff		•				•

Source: U.S. Environmental Protection Agency.

Nanotechnology and Water Filters

Several companies are developing nanotechnology-based filters that will clean polluted water. The filters in the treatment system can sift out bacteria, viruses, heavy metals, and organic material.

One filter product consists of spiral-wound layers of fiberglass sheets. The sheets create a permeable surface of nano-sized pores—a bit-like a nanoscale strainer. When pressure is exerted, the water pushes through the pores keeping out viruses and bacteria that are too large to go through the pores.

In Australia, one company has patented a water treatment technology that uses nanoparticles for water purification. The product called MesoLite has been tested in several countries. The treatment can be used to remove ammonia from contaminated wastewaters. Once the ammonia has been extracted in the treatment stage, the leftover ammonia can be recycled and used as a commercial fertilizer. The MesoLite process can also be used as a backup system to support large wastewater treatment plants.

DRINKING WATER FROM THE OCEAN

The need for a sustainable, affordable supply of clean water is a key priority for the United States' future and especially for states that have few freshwater sources and a fast-growing population. One source of providing more fresh water to ocean-side states is to transform unusable water from the ocean into valuable freshwater. Desalination is the process that is used for removing dissolved salts from seawater and brackish water.

Reverse Osmosis

In a water desalination treatment system, reverse osmosis is a separation process that uses pressure to force a solvent, such as water, through a membrane that retains the solute, such as salt ions, on one side and allows pure water to pass to the other side. However, reverse osmosis treatment units use a lot of water and energy and therefore can be costly to operate. As an example, some reverse osmosis treatment plants recover only 5 to 15 percent of pure water entering the system. The remainder is discharged as wastewater. One nanotechnology group may have a way to reduce costs of conventional desalination treatments by using carbon nanotubes.

A New Kind of Reverse Osmosis

Researchers at Lawrence Livermore National Laboratory (LLNL) are developing a water desalination system using carbon nanotube-based membranes. The use of the carbon nanotubes in the membranes could reduce the cost of desalination by 75 percent, compared to reverse osmosis methods used today.

The carbon nanotubes used by the researchers are sheets of carbon atoms rolled so tightly that only seven water molecules can fit across their diameter. Their small size makes them a good source for separating molecules. The nanopores also reduce the amount of pressure needed to force water through the membrane As a result, there are savings in energy costs when compared to reverse osmosis using conventional membranes.

Water Pollution: Using the Atomic Force Microscope to Study Water Pollution

Virginia Tech researchers are using an atomic force microscopy (AFM) to observe how a bacterium adheres to the silica surfaces of soil. This sticking efficiency of bacteria has not been previously measured experimentally using the AFM. The researchers believe if they can learn how bacteria can adhere or stick on to various soil surfaces, then they can use the information to predict how bacteria in the soil particles can be removed and transported in the groundwater.

GROUNDWATER POLLUTANTS

In the United States, approximately 50 percent of the people depend on underground water naturally stored in aquifers. Surface water provides the other source of fresh water. In fact, many of the rural areas of the United States depend almost entirely on groundwater. But some of the nation's groundwater is contaminated, say scientists, and clean up could cost hundreds of billions of dollars as well as several decades to complete.

Groundwater occurs beneath Earth's surface at depths of a few centimeters to more than 300 meters (900 feet). The water that is available for human use by pumping operations is within the zone of saturation. The zone of saturation is where the spaces between particles of soil or spaces within fractures of rock that compose the aquifer are entirely filled with groundwater. Groundwater concerns include the leaching of pollutants such as arsenic and MTBE (a gasoline additive, now banned) into the water making it unfit for human consumption. The leaching of buried toxic and hazardous wastes can also pollute groundwater resources.

GROUNDWATER CLEANUP

Scientists are developing different technologies for cleaning up pollutants in groundwater.

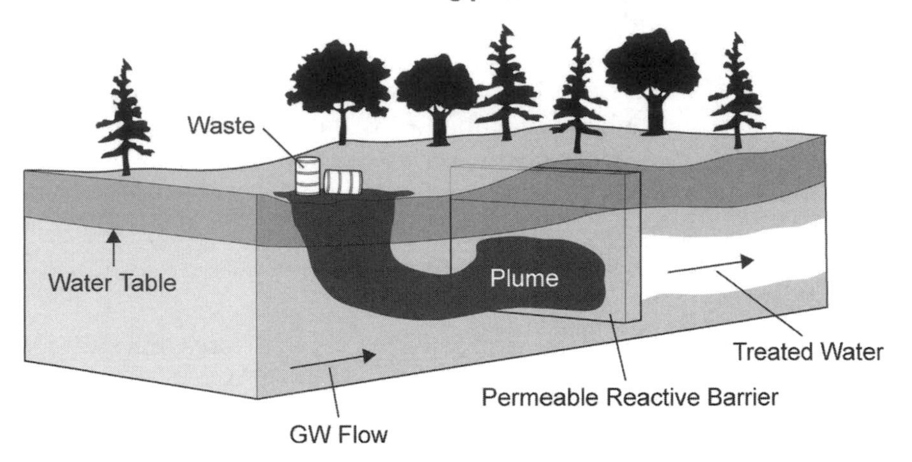

Figure 8.1 Contaminant remediation using permeable reactive barriers. The Permeable Reactive Barriers (PRBs), commonly called "iron walls," is another approach to remediate contaminated groundwater. In this process, macro-sized iron particles are placed into the ground forming a permeable iron barrier or wall that dissolves the pollutants in the groundwater. (*Courtesy of Jeff Dixon. Redrawn with permission from Robert M. Powell, Horizon Environmental Corporation*)

Iron Nanoparticles May Be Effective in Cleaning Up Waste Sites

Professor Zhang of Lehigh University has been working for several years to develop a remediation method to clean up groundwater contaminants, using nanoscale metallic particles. He has been field-testing the method at several industrial sites. The industrial sites are contaminated with such toxicants as polychlorinated biphenyls (PCBs), DDT (a potent pesticide), and dioxin. These are all chlorinated hydrocarbons and persistent organic pollutants of types that are fairly insoluble in water; they are found mostly in soil sediments, and fatty animal tissue. So far, the results of his research have been very encouraging.

Professor Zwang's technology involves pumping nanoparticles (iron-based particles measuring 100 to 200 nm) into the contaminated groundwater. The nanoparticles are almost all iron with less than 1 percent palladium. Palladium (Pd) is a rare silver-white metal and is used as a catalyst in catalytic converters on cars. When the iron-based nanoparticles are applied to water or soil contaminated with carcinogenic chlorinated solvents, the particles remove the chlorine and convert the solvents into harmless hydrocarbons and chloride, commonly found in table salt.

Professor Zhang's nanoremediation offers potential advantages over existing conventional methods. One of the advantages is that nanoremediation avoids digging up the contaminated soil and treating it—this is a very costly treatment process.

CLEANING UP ORGANIC POLLUTANTS USING NANOTECHNOLOGY

Trichloroethene or TCE is one of the most common and poisonous organic pollutants in U.S. groundwater and one of the nation's most pervasive and troublesome groundwater pollutants. TCE is an industrial solvent used primarily in metal degreasing and cleaning operations.

TCE is found at 60 percent of the contaminated waste sites on the Superfund National Priorities List, and it is considered one of the most hazardous chemicals at these sites because of its prevalence and its toxicity. Superfund is the common name for the Comprehensive Environmental Response, Compensation and Liability Act (CERCLA) of 1980. This is a federal law whose mission is to clean up the worst hazardous and toxic waste site areas on land and water. These sites constitute threats to human health and the environment.

TCE, as a volatile organic compound (VOC), has a tendency to volatilize or escape from groundwater into the air. When this process of volatilization happens in groundwater beneath a building, the TCE can enter the building as a vapor, which can produce an air quality or inhalation hazard to people within the building.

TCE can be absorbed through the lungs, mucous membranes, gastrointestinal tract, and the skin. Exposure to TCE happens mostly from breathing contaminated air and drinking contaminated water. Short-term exposure to high levels of this chemical can result in toxic effects on a number of organs and systems, including the liver, kidney, blood, skin, immune system, reproductive system, nervous system, and cardiovascular system. In humans, acute inhalation exposure to TCE causes central nervous system symptoms such as headache, dizziness, nausea, and unconsciousness. TCE has been linked to liver damage, impaired pregnancies, and cancer.

TCE Cleanup with Gold Nanoparticles

Cost is the major hurdle to cleaning up TCE-polluted groundwater. Cleanup costs for TCE nationwide are estimated in the billions of dollars.

The Department of Defense alone estimates the cost of bringing its 1,400 TCE-contaminated sites into Environmental PA compliance at more than $5 billion.

Researchers at Rice University and Georgia Institute of Technology have found that gold nanoparticles coated with palladium are extremely effective catalysts for breaking down toxic chemicals such as TCE into less harmful products. Using a treatment of gold nanoparticles, the researchers hope to drive down the costs of cleanup operations by eliminating:

- the drilling costs for new wells,
- the construction costs for surface treatment facilities, and
- the energy costs of pumping up water to the surface.

Tests have found that the gold-palladium nanocatalysts break TCE down about 100 times faster than bulk palladium catalysts. One of the major advantages of using palladium catalysts to break down TCE is that palladium converts TCE directly into nontoxic ethene and ethane, colorless, odorless, and gaseous hydrocarbons.

The researchers have another idea and that is to develop a device that would include a cylindrical pump containing a catalytic membrane of the gold-palladium nanoparticles. The device would be placed down existing wells where it would pump water through continuously breaking the TCE into nontoxic components.

CLEANING UP NUCLEAR WASTE SITES

Scientists believe that using gold-palladium nanoparticles may have an impact on other toxic waste sites, such as the Hanford Nuclear Waste Site located in the southeastern part of Washington State. The site, near the Columbia River, contains nuclear waste material since the end of World War II. Many environmental activists are concerned that the waste will get into the groundwater. So, maybe a new kind of treatment method using nanoparticles could play a role in preventing the spread of the nuclear wastes into the groundwater.

AIR POLLUTION

Air pollution is a major health problem in the United States and throughout the world. An estimated 3 million people die each year from the effects of air pollutants. Medical researchers have linked high levels of air pollution to illnesses and diseases. These health problems

Table 8.2 Applications of Titanium Oxide (TiO_2) as a Photocatalyst

• Self-cleaning and bacterial degradation for glass products and for floors and walls in hospitals.
• Water cleaning capacity and for purification of soil.
• Air cleaning effects for pavement blocks and sidewalks.

include asthma, allergies, emphysema, chronic bronchitis, lung cancer, and heart attacks. Pollutants in the air cost citizens billions of dollars every year in health care and lost time at work. In a major health study conducted by the American Lung Association, about 50 percent of the United States population is breathing unhealthy amounts of air pollution. Other countries suffer from air pollution too.

Titanium Dioxide and Clean Air

Walking down a street in central London, a person may suddenly feel that the air quality is quite clean. The sidewalk has been treated with a nanotechnology product containing catalytic properties that break down molecules in harmful pollutant emissions in the air. In Milan (Italy) and Paris (France), there are similar concrete sidewalks that are treated with titanium dioxide (TiO_2). According to some experts, the titanium dioxide concrete slabs have reduced pollutants during rush hour by 60 to 70 percent in these two cities.

Japan's Mitsubishi Materials Corporation has developed a paving stone that uses the catalytic properties of TiO_2 to remove nitrogen oxides (NO_x) from the air. Nitrogen oxides are emitted into the air primarily from the emissions of automobiles and power plants that burn petroleum and coal. The paving stone breaks down nitrogen oxides into more environmental-friendly substances such as nitric acid ions. These ions can then be washed away by rainfall or neutralized by the alkaline composition of the concrete.

What Is Titanium Dioxide?

Titanium dioxide is a white powder. It is used as a whitener for paints, in food products, toothpaste, and many other consumer goods. Titanium dioxide is a strong photocatalyst. A photocatalyst produces surface oxidation to eliminate bacteria and other materials when it is exposed to the sun or a fluorescent lamp. Therefore, when a product is made of TiO_2 in the sun, organic compounds such as dirt, mold and mildew, and bacteria break down, without consuming the TiO_2

Another advantage of products made from titanium dioxide is that it has a self-cleaning surface. When water hits the titanium dioxide, a smooth sheet forms instead of tiny droplets. The sheet of water gets under the dirt and lifts it off the surface and washes it away. The combination of these effects makes TiO_2 self-cleaning to a large extent, and it can be applied in microscopically thin coatings to building materials, glass, or even fabrics.

Scientists in Hong Kong are developing dirt-resistant clothing with titanium dioxide and a Japanese company has manufactured ceramic tiles using the chemical. The titanium dioxide self-cleaning ceramic tiles, according to the National Institute of Health's Environmental Health Perspectives, have achieved a 99.9 percent bacterial kill rate within one hour for such strains as penicillin-resistant *Staphylococcus aureus* and *Escherichia coli.*

Did you know?

Not all air pollution occurs outdoors. In fact, indoor air is often even more polluted than the air outside homes and workplaces. The United States Environmental Protection Agency (EPA) Science Advisory Board has ranked indoor air pollution among the top five environmental risks to public health. Many of these indoor air quality hazards are attributable to VOCs, such as TCE and other organic solvents.

ENVIRONMENTAL PROTECTION AGENCY AND DEPARTMENT OF ENERGY

Two major federal agencies that will be involved in the development, funding, and applications of nanotechnology for environmental programs to conserve and protect resources and to provide new sources of energy will be the Environmental Protection Agency (EPA) and the Department of Energy (DOE).

Environmental Protection Agency

The Environmental Protection Agency was created in 1970 and was established in response to growing public concern about: unhealthy air, polluted rivers and groundwater, unsafe drinking water, endangered species, and hazardous waste disposal.

The agency's mission is to "protect public health and to safeguard the natural environment—air, water, and land—upon which human life depends." Its areas of responsibility include control of air pollution and water pollution, solid waste management, protection of the drinking water supply, and pesticide regulation. The EPA is aware that nanotechnology is a revolutionary science and engineering approach that has the potential to have major consequences—positive and negative—on the environment.

Department of Energy

The Department of Energy's (DOE) goal is to advance the national, economic, and energy security of the United States. The department's strategic goals to achieve the mission are designed to deliver results along strategic themes that include:

- Energy Security: Promoting America's energy security through reliable, clean, and affordable energy,
- Environmental Responsibility: Protecting the environment by providing a responsible resolution to the environmental legacy of nuclear weapons production.

NANOTECHNOLOGY AND ENERGY SOURCES

In this century, along with renewal energy sources such as wind energy, geothermal energy, and nuclear energy, many companies are focusing on the technology to produce photovoltaic solar cells and hydrogen fuel cells.

Solar Photovoltaic Cell (PV)

A solar photovoltaic cell (PV) is a device that converts solar energy into electricity in a manner that does not release any pollutants to the environment. Today, solar cells are commonly used to power small-sized items such as calculators and watches. But solar PVs have a great future in providing all the electricity needs for rural communities, homes, and businesses. The future for solar cell production and their usage is very promising as a renewable energy resource. However, the technology still remains expensive when compared to the costs of fossil fuels to produce electricity.

Solar cells have two major challenges: they cost too much to make (in the form of energy), and they are not very efficient. Although efficiencies of 30 percent have been achieved, the present-day typical efficiencies for solar cells is from 15 to 20 percent. However, solar nano researchers believe they can make solar cells more efficient.

Quantum Dots for Solar Cells

Much of the sun's energy is wasted by today's photovoltaic cells. When solar photons strike a solar cell, they release electrons in the semiconductor to produce an electric current. However, when an electron is set free by the photon, it collides often with a nearby atom making it less likely to set another electron free. So even though the sun's solar photons carry enough energy to release several electrons, producing more electricity, they are limited to one electron per solar photon. As a result, conventional solar cells operate at 15 to 20 percent efficiency using solar energy.

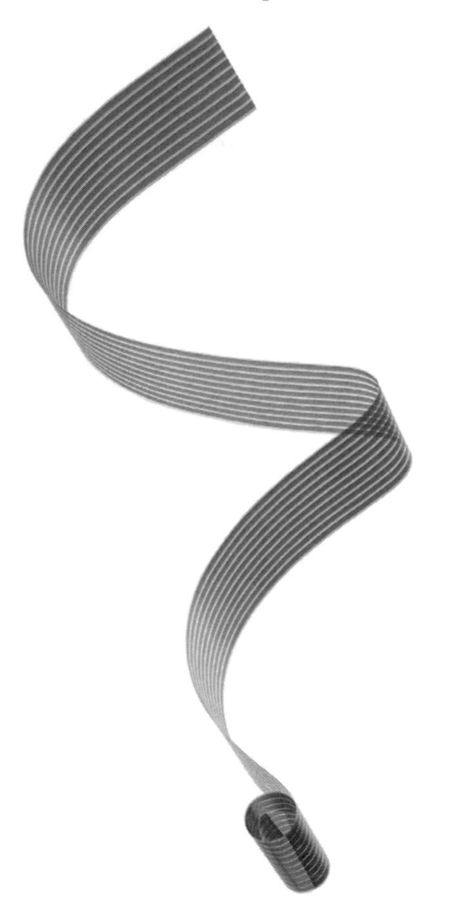

Figure 8.2 Konarka Technologies develops light-activated Power Plastic® that is flexible, lightweight, low in cost and much more versatile in application than traditional silicon-based solar cells. These new materials are made from conducting polymers and nano-engineered materials that can be coated or printed onto a surface in a process similar to the way photographic film is made. (*Courtesy Konarka Technologies, Inc.—Power Plastic*™)

Scientists have been doing a lot of research and experiments with quantum dots to make photovoltaic cells more efficient. Researchers at the National Renewable Energy Laboratory (NREL) and the Los Alamos National Laboratory have been experimenting with quantum dots as a semiconductor in a solar cell. They have discovered that the use of the quantum dots allows solar energy to release multiple electrons, not just one. This research has the potential to make major improvements in the manufacturing of photovoltaic cells. The two research teams have calculated that a maximum of solar conversion to electricity to a 42 percent efficiency rate is possible from the conversion of solar energy to electricity. The solar cells could be used to make hydrogen directly from water for fuel cells. The researchers still need more time and research to complete their studies.

Hydrogen Fuel Cells

The next technological advance in energy will be the use of fuel cells. Hydrogen fuel cells are devices that directly convert hydrogen into electricity. Fuel cells can provide electricity to power motor vehicles and to heat and light homes, office buildings, and factories.

The hydrogen fuel cell vehicle is also an excellent alternative to fossil fuel vehicles because hydrogen produces no carbon dioxide when burned, and the fuel cell requires little maintenance because it has few moving parts.

Fuel Cells and the Automobile Industry

Presently, carmakers are investing billions in hydrogen fuel cell research and are testing prototype fuel-cell vehicles. Within the next decade or so, some automakers and other experts believe the hydrogen fuel cell will replace the need for petroleum, diesel, and natural gas as the main fuel for automobiles, buses, and trucks.

> **Did you know?**
> Fuel cells are used in NASA's space program to provide heat, electricity, and drinking water for astronauts. Fuel cells are also used aboard Russian space vehicles.

There are many different types of hydrogen fuel cells that use different electrolytes, operate at different temperatures, and are suited to different uses. Presently, the most popular fuel cell for automobiles is called the proton exchange membrane (PEM). It is a lightweight fuel cell that is one of the easiest to build. The outside portion of the fuel cell, or membrane, is coated with platinum, which acts as a catalyst. Hydrogen under great pressure and temperature is forced through the catalyst. At this point, the element is stripped of its electrons, allowing them to move through a circuit to produce electricity. The hydrogen protons pass through the membrane and combine with oxygen in the outside air to form water. There is no polluted emission, only water is the by-product.

Today, Hydrogen used in most fuel cells is made from "reforming" methane with high-pressure steam. The steam interacts with the methane to separate the hydrogen from the hydrocarbon molecules in the methane. Special equipment, called reformers, can

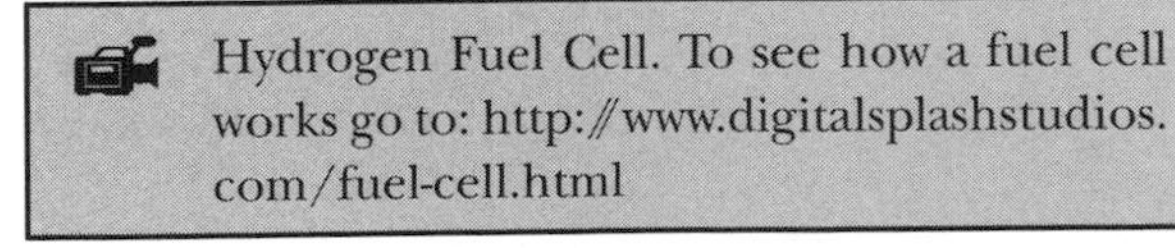
Hydrogen Fuel Cell. To see how a fuel cell works go to: http://www.digitalsplashstudios.com/fuel-cell.html

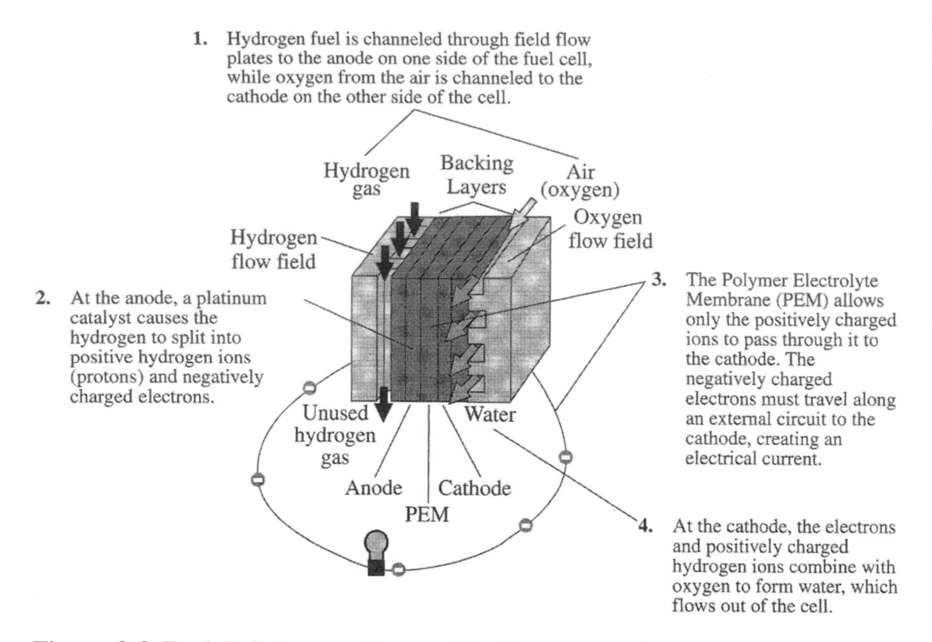

Figure 8.3 Fuel Cell Source. *Source: U.S. Department of Energy.*

make hydrogen from coal, fossil fuels, nuclear energy, sewage, bacteria, and paper-mill waste. In the future, solar cells may be used to make hydrogen.

The Problem with Fuel Cells

One of the major obstacles in fuel-cell research is designing a safe, lightweight, and compact hydrogen fuel tank that can store sufficient amounts of hydrogen.

Hydrogen is a gas, and is not dense, so to store enough of it in a tank you have to compress it at very high pressures. Another option is to store the hydrogen gas chemically by bonding it with another material. The carbon nanotubes maybe the answer to the storage problem. The reason is that hydrogen easily bonds to carbon nanotubes. Placing a grid of coated nanotubes (or buckyballs) inside the tank, the nanotubes would "soak

Did you know?

The first fuel cell was built in 1839 by Welsh judge and scientist Sir William Grove. However, the use of fuel cells as a practical generator began in the 1960s, when the National Space Administration (NASA) chose fuel cells over nuclear power and solar energy to power the *Gemini* and *Apollo* spacecraft. Today, fuel cells provide electricity and water on the space shuttles.

up" the hydrogen like a sponge. As a result, the nanotube grid has the capacity to absorb large amounts of hydrogen gas in a tank about the size of an automobile gas tank.

Once the tank is filled up with the hydrogen, the driver would start the car engine. This action would cause the hydrogen to dislodge from its storage area, the tank, and float through a hose into the fuel cell. In the fuel cell, the hydrogen would be converted into electricity and water vapor. However, while fuel cells look promising, more research will need to be done before a mass production of hydrogen fuel cell vehicles are on the road.

EMERGING NANOTECHNOLOGIES AND RISKS

Industry and academic experts suggest there is still uncertainty about the impact of releasing nanoparticles into the environment. It has also been pointed out that there is a lot more to learn about how nanoparticles behave when they are released into the environment. Like all emerging technologies with great promise, the nanotechnology and nanochemistry industries will face challenges in ensuring that environmental risks are properly managed.

NANO INTERVIEW: PROFESSOR PAUL G. TRATNYEK, Ph.D., OREGON HEALTH & SCIENCE UNIVERSITY'S OGI SCHOOL OF SCIENCE & ENGINEERING

Professor Paul G. Tratnyek is an environmental chemist and a professor of environmental and biomolecular systems at Oregon Health & Science University's OGI School of Science & Engineering, Beaverton, Oregon.

Dr. Tratnyek's research group, along with collaborators at the Pacific Northwest National Laboratory and the University of Minnesota, has discovered that at least one type of nano-sized particles of iron may be helpful in cleaning up carbon tetrachloride contamination in groundwater. Carbon tetrachloride is a chemical that is used in dry-cleaning applications and as a degreasing agent to clean tools. Carbon tetrachloride is a toxic chemical that has shown to cause cancer in animals.

The prospect of using iron nanoparticle technology for environmental cleanup is promising. However, Dr. Tratnyek states that there are a lot of unanswered questions about the appropriate implementation of this technology and even some questions about its safety.

Dr. Tratnyek talked to the author about his career and his environmental work in groundwater contamination.

Where did you grow up and what schools did you attend?

I grew up in Sudbury, Massachusetts, a suburb of Boston. I went to the public high school and then to Williams College in Western Massachusetts where I received my bachelor's degree. After Williams College, I went to the Colorado School of Mines in Colorado and received my Ph.D. degree in applied chemistry. The reason I selected the Colorado School of Mines to study was because I was interested in the link between chemistry and the earth sciences.

Professor Paul G. Tratnyek is an environmental chemist and a professor of environmental and biomolecular systems at Oregon Health & Science University's OGI School of Science & Engineering, Beaverton, Oregon. (*Courtesy Anne Rybak Photographic*)

What were some of your favorite activities and subjects in school?

As a young person in elementary school I was very interested in the environmental science activities and natural history. Through high school and college I focused mainly on the "basics," i.e., chemistry, physics, and mathematics. Only in graduate school did I come back around to studying environmental science.

How did you get interested in using iron particles in cleaning up contaminants in groundwater?

When I arrived at the Oregon Graduate Institute and became a professor, one of my senior mentors, another professor in the department, knew the group in Canada that had first stumbled onto the possibility of using iron metal to remediate contaminated groundwater. But the group had difficulty addressing the chemical aspect of that technology. I offered to help with that, and they provided a small amount of funding for me to get started. My early results helped explain how the technology worked and this gave people confidence and boosted the commercial viability of the technology.

How does groundwater become contaminated?

Groundwater contamination occurs all over the place and comes from two forms—point sources and nonpoint sources. Point source means that the contamination was discharged to the environment from some localized source, such as a dry-cleaning plant or from a factory or business that had a leaking pipe or had spilled the contaminants. Nonpoint sources include contamination that was distributed broadly across a wide environment. Pesticides often present nonpoint source contamination problems because they are introduced to the environment over large areas, by aerial spraying, or the like.

Nonpoint and point sources are both big problems and it is difficult at times to know where the source of the contamination came from. What you usually hear in the news is the point source contamination cases because you can get names of some companies who buried toxic materials in their backyards. Many of these are Superfund sites, which get a lot of press.

What are some of the contaminants found in groundwater and which ones are the most harmful to human health?

Overall, the most significant contaminants probably are the chlorinated solvents, which were used in dry cleaning, degreasing, cleaning circuit boards in the electronic industry, etc. Big and small spills of these chemicals are widespread. The chlorinated solvents are toxic (e.g., some are carcinogenic) and therefore present real problems when they contaminate groundwater used for drinking purposes. Besides chlorinated solvents, there are other major categories of contaminants: gasoline and other fuels; metals like chromate, etc.

What is the most common or traditional method that is used for groundwater remediation?

The most traditional method is referred to as a "pump and treat." This involves pumping the groundwater out of the ground and treatment of the contamination at the surface. Because treatment is done above ground, it's called an "ex situ" method.

In this process, a well is put into the ground where there is contaminated groundwater. A pump is attached to the well and if the pumping is successful it will draw some of the contamination out of the ground to the surface. Then the contaminants are treated in a variety of different ways, usually be adsorption to activated carbon filters. The pump and treat is easy from an engineering standpoint. However, this process is also criticized as being inefficient for two related reasons. The pump and treat method tends not to solve the problem, but just controls it. This means that you have to continue to pump these contaminated areas for a long period. If the well pumps are turned off, the contaminant concentrations tend to rebound.

The constant daily pumping operation is costly. Ex situ process can also be tied up in legal and regulatory issues because once the contaminated water is pumped up to the surface you cannot just dump it back on the ground again because it is hazardous waste, so you need to go through regulatory and legal requirements to remove it.

You have been studying how iron particles break down pollutants since 1992. One technology you and your colleges helped to develop was the use of Permeable Reactive Barriers (PRBs), commonly called "iron walls," to remediate contaminated groundwater. In this process, however, you used macro-sized iron particles. What are iron walls and what are the advantages of using this process for cleanup rather than the traditional remediation technologies?

The "iron wall" or permeable reactive barriers (PRBs) has two main characteristics—it is in situ and it is passive. In situ means that you apply iron particles by placing them into the ground and treat the contamination there. You do not have to pump the contaminated water up to the surface. The other advantage is that this method is passive. Pump and treat is active because you have to constantly keep the pumps working. Ideally passive technologies mean that once the treatment is in place you can cover it up, plant grass, and then you can leave the area—there is nothing else to do.

Passive can be much less expensive in the long haul than pump and treat which is an active process. In situ allows you to put iron particles and materials down into the ground to interact with the groundwater to remove the contaminants. The groundwater will flow through this treatment zone and be treated and what comes out at the end is cleaned. The iron wall also serves the function of an in-ground cutoff wall. This means you can intercept the contaminated groundwater in a very precise way—you cut it off. This cutoff wall is useful in many ways where many properties are near one another. So, if someone spills toxic materials into the ground and contaminates the groundwater, you can use a cutoff wall to keep the contaminants from entering the groundwater of the nearby neighbor's property, or discharging into surface water.

How does iron break down the contaminants?
Traditionally iron walls have been made with granular iron particles that are the size of millimeters. These coarse iron grains are fairly inexpensive and plentiful. When contaminated water interacts with the metal, a chemical reaction takes place. The reaction removes the chlorines, which makes the products less toxic. In a traditional iron wall, the iron remains available to degrade contaminants pretty much indefinitely.

Do the iron particles have any advantages over the use of other metals for groundwater remediation?
You can use other metals but iron is better because it is less harmful to the environment than other metals. Iron is cheaper too. There are two methods of using iron particles in groundwater remediation.

The traditional way or the most common way of making permeable iron walls (PRBs) is to use iron particles that are manufactured for the construction industry. These micron-sized iron particles are coarse and granular-shaped.

Recently, however, there has been interest in the PRB field in using nano-sized iron particles for cleaning up contaminants in groundwater. Nano-sized particles range in size from 10 to 100 nanometers. Much of this process is similar to the conventional iron wall method for clean up. However, there are a few differences. The nanoparticles particles are so small that you can make slurry solutions and then inject the solutions into the wells. By using this process, you can create an iron wall, but at a much greater depth than the conventional PRB way. This is huge advantage. The downside is that when you inject the iron nanoparticles into the ground, it is not clear how far the particles will move once they are in the groundwater.

Generally, nanoparticles do not move very far, in fact they do not move far enough. You want the nanoparticles to move farther. Researchers are working on ways to make nanoparticles go farther, but if they get too good at this the particles might move to places we don't want them (drinking water supplies, for example). So, we need to do more research to try to figure out if this PRB process poses a potential problem.

Once these "iron walls" are placed in the soil how long is it before they need to be replaced?
The installation of the conventional non-nano iron wall has lasted for some time. There are a good number of iron walls that have worked for more than 10 years.

We expect that the nanoparticle iron wall installation will not last very long, because the nanoparticles are so small that they will be used up in a short time. Therefore, it is difficult to estimate how long the nanoparticle iron wall will last. Our best estimate is that the nanoparticle iron wall will probably last only for a few months. The goal of the nanoparticle iron wall is to knock out the contaminants quickly.

Where have these "iron walls" been used in Oregon and elsewhere?
The convention iron wall (PRBs) was first used by researchers at the University of Waterloo at a site in Ontario, Canada. In the U.S., the first large-scale conventional iron wall was used in an industrial park in Sunnyvale, California. Since then, there have been many applications of the conventional PRB.

The iron nanoparticle remediation process is much newer and there are fewer of those that have been implemented. The first pilot program was initiated by Professor W.X. Zhang at Lehigh University. He established a pilot program at a chemical manufacturing site on the east coast and reported some favorable results. This was the first pilot-scaled program and since then there have been a number of consulting companies that have done more than a dozen versions of the nanoparticle iron wall installations.

Are there any risks in using iron nanoparticles for groundwater remediation and if so, what would some of these risks be?
There is much debate going on about the use of nanoparticles. The risks have two parts; one is toxicity and the other part is the exposure part. Nanoparticles can be toxic, especially if you inhale them, but will there be any harm to humans and animals? That depends on whether there is exposure to these iron nanoparticles.

Different situations pose different risks. If you are in the business of manufacturing these particles then you have occupational health issues that you need to worry about. However, once you inject these particles into the ground for remediation there is very little chance that anyone will be exposed to those iron particles. It is quite clear that these particles are not very mobile in groundwater because they tend to stick onto anything available for them to attach to. A general rule is once you inject these particles into the ground they may move only a few feet from where you have injected them. So there is little likelihood that the nanoparticles will show up in the neighbor's drinking water well that is 300 feet away.

I heard that scrap iron is used to make iron particles used for groundwater remediation. Is this true? If so, how is it used?
The conventional commercial granular iron micro-particles (not iron nanoparticles) that are used in a conventional iron wall come from a

variety of scrap metal dealers, many from the Midwest. These materials come from sources like old brake drums. The scrap iron is shredded and then it is cleaned to remove any grease or oil on the metal, and is sold for a variety of commercial applications such as hardening concrete.

The iron nanoparticles are not made this way. Iron nanoparticles are usually made in a process called bottom-up nanofabrication. Nanoparticles start with a solution of iron and some other materials and then you precipitate the nanoparticles out of the solution starting from the molecules going to the nanoparticles.

In one of your articles, you described an experiment for undergraduate general chemistry students. In the experiment, the students investigated the chemistry behind the iron-permeable reactive barriers (iron walls) used to remediate contaminated groundwater. Is this an activity that high school chemistry students could also perform safely, under supervision, in their science class?
Yes, Students and teachers can refer to a 4-page article, *A Discovery-Based Experiment Illustrating How Iron Metal is Used to Remediate Contaminated Groundwater.* For the article, go to: http://cgr.ebs.ogi.edu/merl/resources.htm

From reading the article, teachers and students can get enough information to investigate the details of the degradation process in the PRB contaminant system by determining the various parameters for the degradation of a dye in place of a contaminant. Dyes have often been used as model contaminants. So, we use a dye in this activity because it is nontoxic and it works well.

In the activity, students measure the kinetics of the disappearance of a dye solution in a cuvette containing granular iron particles—not the iron nanoparticles. There is a nice illustration in the article that shows the degradation process of the dye. If you wish, you can also monitor the dye concentration by using a UV-vis spectrophotometer. We have had high school teachers who have worked on several variations of this activity.

Resources for Educators and Students

1. A multimedia CD-ROM (*MERL: Metals for Environmental Remediation and Learning*), which provides more details on the history, development, installation, and chemistry of iron walls, is available from the authors. See http://cgr.ebs.ogi.edu/merl/ (accessed on October 2001) for more information.

2. Useful information on iron walls can be found at many Web sites, including http://cgr.ebs.ogi.edu/iron; http://www.rtdf.org; http://www.doegjpo.com/perm-barr/.

NANO ACTIVITY: DISCOVERING THE PROPERTIES OF FERROFLUIDS (Teacher Supervision Advised)

Nanotechnologists are experimenting with ferrofluids to help clean up wastewater. Ferrofluids contain nanoparticles of a magnetic solid, usually magnetite (Fe_3O_4), in a colloidal suspension.

NASA discovered ferrofluids when their scientists were trying to control liquid products in space. Ferrofluids have been used in many applications, including computer disk drives, low-friction seals, and loudspeakers.

In this activity, you will learn about some of the properties of ferrofluids. The activity was adapted from the *NanoSense* Program produced by the SRI International in Menlo Park, California. Patricia Schank and Tina Stanford coauthored the activity. To learn more about this nano activity and other *NanoSense* activities go to: www.nanosense.org (Safety Note: You should do the following activity with teacher supervision.)

Materials You Need

1. Sealed display cells of ferrofluids (for ordering, see Where to Buy)
2. Magnet

- Have your teacher show and pass around one or more samples of ferrofluid along with a strong magnet. Under teacher supervision, let students play with the ferrofluids and magnet and see what they can make it do. The teacher may also want to show and pass around another magnetic material, like a piece of iron, for comparison. The teacher should explain to the students that since we have been able to make the particles in the ferrofluids so small, we have been able to change the physical state of the material from a solid to a liquid.
- What happens when you bring a magnet close to the liquid?
- When you put the magnet near the ferrofluids, it distorts. What causes this distortion?
- What does this distortion represent?

Where to Buy Ferrofluids

The science department or the science teacher can order sealed display cells of ferrofluids from Educational Innovations, Inc., at: http://

www.teachersource.com (click on "Browse or Search the Catalog", "Electricity! Magnetism! Engines!" and then "Ferrofluids"). A Ferrofluid Experiment Booklet is also available (item FF-150) for about $6.

References

http://jchemed.chem.wisc.edu/JCESoft/CCA/CCA2/MAIN/FEFLUID/CD2R1.HTM

READING MATERIAL

Karn, Barbara, Vicki Colvin, Paul Alivasatos, Tina Masciangioli (Eds.). *Nanotechnology and the Environment.* Washington, DC: American Chemical Society, 2004.

Lewis, Scott Alan. *The Sierra Club Guide to Safe Drinking Water.* San Francisco: Sierra Club Books, 1996.

Roco, Mihail C., and William Sims Bainbridge. *Societal Implications of Nanoscience and Nanotechnology.* Boston, MA: Kluwer Academic Publishers, 2001.

Somerville, Richard C.J. *The Forgiving Air: Understanding Environmental Change.* Berkeley, CA: University of California Press, 1998.

Wiesner, Mark, and Jean-Yves Bottero. *Environmental Nanotechnology.* New York. McGraw Hill Professional Publishing, 2007.

VIDEOS

Photovoltaics. Turning Sunlight into Electricity. United States Department of Energy. Solar Energies Technologies Program: Animations. The "photovoltaic effect" is the basic physical process through which a PV cell converts sunlight into electricity. Sunlight is composed of photons or particles of solar energy. http://www1.eere.energy.gov/solar/video/pv3.mov

Clean Technology Vehicles. Altairnano. Demonstration vehicle using NanoSafe long-term batteries for vehicles. http://www.altairnano.com/ZEV.mov

Hydrogen Fuel Cell. Digital Splash Multimedia Studios. http://www.digitalsplashstudios.com/fuel-cell.html

How a Fuel Cell Works. Ballard Fuel Cells. Ballard's principal business is the design, development, and manufacture of proton exchange membrane (PEM) fuel cell products. (Scroll down)

http://www.ballard.com/be_informed/fuel_cell_technology/how_the_technology_works

WEB SITES

National Institute for Occupational Health (NIOSH): http://www.cdc.gov/niosh/homepage.html

Copies of the *National Air Quality and Emissions Trends Report* are available from U.S. Environmental Protection Agency. Copies of the report may be viewed via the OAR Web site at: http://www.epa.gov/docs/oar/oarhome.html

Environmental Protection Agency: http://www.epa.gov/

A site for innovative technologies for contaminated soil and groundwater.

Department of Energy: http://www.energy.gov/index.htm

National Renewable Energy Laboratory (NREL): www.nrel.gov/

Sandia National Laboratories: http://www.sandia.gov/

SOMETHING TO DO

Do you ever wonder how much energy your school building uses? Could you become an energy consultant who could design a nanotechnology program to save energy? If you do, you may want to go to the EnergyNet sponsored by the Department of Energy. They have several energy projects on their Web site.

Energy Audits

Here is one online project from the Department of Energy that will give you and others in grades 6–12 the opportunity to act as energy consultants for their schools. In this project, you can conduct energy audits to review the energy use in your school or other buildings. For directions go to:

http://www.energynet.net/eninfo/info/what_is_energynet.html

The U.S. Department of Energy also has a Web site of many other kinds of activities to do. Here is their Web site:

http://www1.eere.energy.gov/education/lesson_plans.html

9

Nanotechnology Projects and the United States Government

Nanotechnology is expected to have a large impact on many sectors of the United States and the world's economy. All governments agree that a strong nanotechnology economy can lead to new products, new businesses, new jobs, and even new industries for many countries. As a result, nanotechnology funding for research is growing rapidly in many countries.

The U.S. government is aware of the economic and social impacts that nanotechnology can make. For several decades, the government has funded nanotechnology research.

NATIONAL NANOTECHNOLOGY INSTITUTE

However, a big push for more government investment in nanotechnology research occurred in 2001. At that time, President Clinton requested in the 2002 federal budget a major new initiative, called the National Nanotechnology Initiative (NNI). The budget included an increase of more than 200 million dollars for the government's investment in nanotechnology research and development.

Additional nanotechnology dollars were authorized during President Bush's term. In December 2003, President Bush signed the Nanotechnology Research and Development Act, which authorizes funding for nanotechnology research and development (R&D) nanotechnology. This legislation puts into law programs and activities supported by the National Nanotechnology Initiative (NNI). The legislation also authorizes public hearings and expert advisory panels, as well as the American Nanotechnology Preparedness Center to study the emerging technology's potential societal and ethical effects.

According to one report, from 2001 to 2007, NNI participants formed over 60 facilities or institutes involved in nanotechnology research

Table 9.1 United States Government Expenditures in Nanotechnology (in millions of dollars, approximately)

Agencies	2002	2003	2004	2005	2006	2007
National Science Foundation	$ 204	$ 221	$ 256	$ 338	$ 344	$ 373
Department of Defense	$ 224	$ 322	$ 291	$ 267	$ 436	$ 345
Department of Energy	$ 89	$ 134	$ 202	$ 216	$ 207	$ 258
Other Federal Agencies	$ 136	$ 142	$ 183	$ 217	$ 251	$ 259

and development throughout the United States. New centers of nanotechnology expertise, most of which are associated with universities, train workers and help educate researchers and developers of technology.

Spending on nanotechnology research and development (R&D) has amounted to several billions of dollars since the late 1990s, by the United States alone and accounts for a large portion of worldwide spending on this type of work. The government agencies that have spent the greatest amounts on nanotechnology are the National Science Foundation, the Department of Energy, and the Nation Institutes of Health.

In nanotechnology research, engineers, medical doctors, educators, chemists, physicists, nanotechnology researchers usually collaborate on studies and projects. Nanotechnology researchers often must share equipment, because many of the high-technology microscopes and other devices used in this technology are too expensive for each individual research facility to purchase and maintain its own. For example, acquiring a scanning electron microscope and building a cleanroom can cost millions of dollars. So, grants and loans from government and private institutions provide money for such activities and equipment.

So far, the NNI has already made valuable contributions to the development of nanotechnology. With NNI funding, researchers have been working on gold nanoshells that can target the destruction of malignant cancer cells (See Chapter 5), low-cost solar cells and quantum dots for energy sources (See Chapter 6), and nanoscale iron particles that can reduce the costs of cleaning up contaminated groundwater (See Chapter 8).

THE NATIONAL NANOTECHNOLOGY INITIATIVE (NNI) AND FEDERAL AGENCIES

The main goal of the National Nanotechnology Initiative (NNI) is to coordinate all of the federal agencies' efforts in nanoscale science, engineering, and technology. Three of these federal agencies were covered in earlier chapters. They included the Department of Agriculture in Chapter 7 and the work of the Environmental Protection Agency and the Department of Energy in Chapter 8.

This chapter will report on nanotechnology research at the National Science Foundation, Department of Defense, National Institute of Standards and Technology, National Institute of Occupational Safety, Department of Homeland Security, and the National Aeronautics and Space Administration.

NATIONAL AERONAUTICS AND SPACE ADMINISTRATION (NASA)

President Dwight D. Eisenhower established the National Aeronautics and Space Administration (NASA) in 1958; this was the time of the then Soviet Union's launch of the first artificial satellite. NASA grew out of the National Advisory Committee on Aeronautics, which had been researching flight technology for more than 40 years. NASA's mission is to pioneer the future in space exploration, scientific discovery, and aeronautics research.

In Chapter 5, you read a little bit about NASA's lab-on-a-chip.

Lab-on-a-Chip

According to NASA, the lab-on-a-chip technology can be used for new tools to detect bacteria and life forms on Earth and other planets and for protecting astronauts by monitoring crew health and detecting microbes and contaminants in spacecraft.

On Earth, some basic lab-on-a-chip technology approaches are being used for commercial and medical diagnostic applications. As an example, the lab-on-a-chip technology can be used as an in-office test for strep throat, or modern in-home pregnancy tests. The hand-held portable device, containing a simple chip design, can conduct diagnostic tests and record test results in a short time for the patient.

Did you know?
Since the lab-on-a-chip is small device, a large number of them could be carried on a Mars rover to search for life and for monitoring microbes inside Martian habitats.

Lab-on-a-chip. The lab-on-a-chip is a microfabricated device that performs chemical and biochemical procedures under computer control, using miniscule quantities of samples to be analyzed. (*Courtesy U.S. Department of Energy, Oak Ridge National Laboratory*)

Spacecraft

In the near future NASA may use combinations of plastics and other materials that will greatly reduce the weight of a spacecraft. Less weight in the spacecraft can reduce launch costs.

Carbon fiber technology has already been used to replace many spacecraft components. Presently, the B-2 Stealth Bomber uses carbon fiber materials in its wings. Carbon fiber composites, for example, are five times stiffer than steel for the same weight allowing for much lighter structures. In addition, carbon fibers have the highest thermal conductivity; they do not overheat. This property allows the carbon fibers to be used as heat dissipating elements on the outside of the craft for spacecraft reentry protection.

A Self-Repair Spacecraft

Using advanced nanotechnology, the spacecraft may adapt to conditions in space travel by rebuilding itself, as needed, while in flight. Solar energy from the spacecraft solar panels would power the computers and assemblers. This would also allow general repair and maintenance to occur without using crew repair astronauts.

Spacecraft Recycling

Recycling aboard the spacecraft will be greatly improved by nanotechnology. Recycling at the atomic level will be very efficient and, in closed environments, such as space stations, this will be crucial. Nanotechnology should also be able to recycle the air efficiently as well, providing a high-quality life support system. Recycling water is also well within the capabilities of nanosystems with all waste molecules being recycled and used elsewhere.

Improvements in NASA Spacesuits

The new nano-designed spacesuits will be light, thin, comfortable, and easy to work with but they will have enhanced strength. The suit will automatically adjust to the contours of the astronaut's body when it is put on. The suit will be able to repair itself.

NASA Space Elevator

A space elevator is essentially a long cable extending from the Earth's surface to a tower in space. The space elevator will orbit Earth at about 36,000 kilometers in altitude.

Four to six "elevator tracks" would extend up the sides of the tower and cable platforms, at different levels. These tracks would allow electromagnetic vehicles to travel at speeds reaching thousands of kilometers-per-hour. The vehicles traveling along the cable could serve as a mass transportation system for moving people, payloads, and power between Earth and space. Security measures are also planned to keep the cable structure from tumbling to Earth.

> **Did you know?**
> What will be the first major space commercial businesses? The top two businesses may be passenger flights and mining. Advertising is also a possibility.

Carbon Nanotubes and the Space Elevator. To build such a sky-high elevator, very strong materials would be needed for the cables (tethers) and the tower. Carbon nanotubes could be the answer for constructing the cables. Carbon nanotubes appear to have the potential strength the space elevator needs.

According to NASA researchers, carbon nanotubes are 100 times stronger than steel. The space elevator will become a realistic possibility with nanotube fibers that nanotechnology will provide. Maximum stress is at the highest point—the altitude, so the cable must be thickest there and taper exponentially as it approaches Earth. Using this material, a cable could be constructed, probably downwards from the space station.

Space Elevator. Scientists envision a space elevator based in the Pacific Ocean. Go to: http://www.sciencentral.com/articles/view.php3?article_id=218392162&language=english

However, NASA states that the space elevator construction is not feasible today, but it could be toward the end of the 21st century.

NATIONAL SCIENCE FOUNDATION

The National Science Foundation (NSF) is an independent federal agency. Their goal is to promote the progress of science; to advance the national health, prosperity, and welfare; to secure the national defense. NSF funds support basic research conducted by America's colleges and universities. About 1,300 projects involving more than 6,000 faculty and students are supported each year.

The NSF funds the National Nanofabrication Users Network (NNUN) and the Network for Computational Nanotechnology (NCN). The NNUN includes five university-based research hubs which are focused on electronics, biology, advanced materials, optoelectronics, and nanoscale computer simulation. The NCN, centered at Purdue University, is linking theory and computation to experimental work that helps turn the promise of nanoscience into new nanotechnologies.

The National Science Foundation funds a number of projects. One example of a funded nanoproject includes nano research work done at the University of Akron. Scientists at the University of Akron have shown how to create a densely packed carpet of carbon nanotubes that functions like an artificial gecko foot—but with 200 times the gecko foot's gripping power. Potential applications include dry adhesives for microelectronics, information technology, robotics, space, and many other fields.

Another NSF program that was funded included work at the University of Wisconsin-Madison. Researchers were able to begin the process of developing nanoscale electronic devices that can be directed to

The National Science Foundation funded an exhibit called "It's a Nano World." The traveling exhibit, organized by Sciencenter in Ithaca, New York, is a hands-on interactive exhibition that introduces children and their parents to the biological wonders of the nanoworld. At this station in the exhibit, visitors are using magnifying glasses to look at tiny parts of feathers, shells, and seeds. (*Courtesy Sciencenter, Ithaca, NY*)

self-assemble themselves automatically. This research development would allow manufacturers to mass-produce nanochips having circuit elements only a few molecules across. The mass production of the nanochips would provide large quantities of chips to be developed with possible lower costs per unit.

National Science Foundation Classroom Resources

The NSF provides Classroom Resources, a diverse collection of lessons and Web resources for classroom teachers, their students, and students' families. Go to: http://www.nsf.gov/news/classroom/

NATIONAL INSTITUTE OF STANDARDS AND TECHNOLOGY (NIST)

The National Institute of Standards and Technology (NIST) mission is to promote U.S. innovation and industrial competitiveness by advancing measurement science, standards, and technology in ways that enhance economic security and improve our quality of life. Researchers in NIST's develop measurements, standards, and data crucial to private industry's development of products for a nanotechnology market that could reach $1 trillion during the next decade. NIST's work also aids other federal agencies' efforts to exploit nanotechnology to further their missions, such as national security and environmental protection.

THE FOOD AND DRUG ADMINISTRATION (FDA)

The Food and Drug Administration (FDA) is an agency of the United States Department of Health and Human Services and is responsible for regulating

- food,
- dietary supplements,
- drugs and cosmetics,
- biological medical products,
- dietary supplements,
- color additives in food,
- blood products, and
- medical devices.

FDAconducts research in several of its Centers to understand the characteristics of nanomaterials and nanotechnology processes. Research interests include any areas related to the use of nanoproducts that FDA needs to consider in the regulation of these products. As an example of current research, FDA is collaborating with other agencies on studies that examine the skin absorption and phototoxicity of nano-sized titanium dioxide and zinc oxide preparations used in sunscreens. If any risks are identified, which may occur from these nano chemicals used in sunscreens, then additional tests or other requirements may be needed.

FDA only regulates certain categories of products. Existing requirements may be adequate for most nanotechnology products that we will regulate. These products are in the same size-range as the cells and molecules that FDA reviewers and scientists associate with every day.

DEPARTMENT OF DEFENSE (DOD)

The mission of the Department of Defense is to provide the military forces that are needed to deter war and to protect the security of the United States. The department's headquarters is at the Pentagon.

The DOD has a long history of supporting innovative nanotechnology research efforts for the future advancement of warfighter and battle systems capabilities. The DOD includes the Defense Advanced Research Projects Agency (DARPA), Office of Naval Research (ONR), Army Research Office (ARO), and the Air Force Office of Scientific Research (AFOSR).

Since the mid-1990s, DOD identified nanoscience as a strategic research area that would require a substantial amount of basic research funding on a long-term basis. The potential of nanotechnology applications will impact many areas for future warfighting. Some of these areas include chemical and biological warfare defense, a reduction in the weight of warfighting equipment, high-performance information technology, and uninhabited vehicles and miniature satellites.

DEPARTMENT OF HOMELAND SECURITY

The Department of Homeland Security is a federal agency whose primary mission is to help prevent, protect against, and respond to acts of terrorism on U.S. soil. The Department of Homeland Security has the capability to anticipate, preempt, and deter threats to the homeland whenever possible, and the ability to respond quickly when such threats do materialize.

As the United States focuses on protecting and defending against terrorism, scientists are conducting research in antiterrorist technology. One of the programs is to develop new kinds of nanoscale sensors to detect explosives and hazardous chemicals at the nanometer level. Most experts agree that sensors are critical to all of the homeland security strategies.

Nanoscale sensors, with their small, lightweight size will improve the capability to detect chemical, biological, radiological, and explosive or CBRE agents. The use of nanoscale sensors for CBRE can be deployed for advance security to

- transportation systems (protection for air, bus, train/subway, etc.);
- military (protection for facilities, equipment and personnel);
- federal buildings (White House, U.S. embassies, and all other federal buildings);
- customs (for border crossings, international travel, etc.);
- civilian businesses; and schools.

Further research will yield sensor technologies that are cheaper and lighter yet far more sensitive, selective, and reliable than current systems.

Underwater Sensor Networks

An area that nanotechnology has much significant promise for is in developing networks of multiple sensors that can communicate with each other to detect very small amounts of chemicals or biological agents. As an example, an underwater sensor network could be used to detect the movement of ships into and out of various seaports. The sensors could also be built into cargo ship containers for detecting any chemical, biological or radiological materials in the cargo.

Did you know?
Nanotechnology-based materials will be essential to fabricate protective garb for emergency response teams and hospital staff who will need to cope with CBRE incidents.

THE MAPLESEED: A NANO AIR VEHICLE FOR SURVEILLANCE

The Defense Advanced Research Projects Agency (DARP) is a central research and development organization for the Department of Defense. The organization is working with an aircraft company to design a surveillance drone shaped like a mapleseed. The remote-controlled nano air vehicles (or NAVs, for short) would be dropped from an aircraft. Then, it would whirl around a battlefield snapping pictures or delivering various payloads. Besides controlling lift and pitch, the wings will also house telemetry, communications, navigation, imaging sensors, and battery power. If built, the NAV may only be about 1.5 inches long and have a maximum takeoff weight of about 0.35 ounces.

Detecting Deadly Chemicals. Science Daily. The Anthrax Scare and a New Tool for Collecting Chemical or Biological Samples. Go to: http://www.sciencedaily.com/videos/2006-12-10/

SAFETY IN NANOTECHNOLOGY: THE NATIONAL INSTITUTE FOR OCCUPATIONAL SAFETY AND HEALTH (NIOSH)

The National Institute for Occupational Safety and Health (NIOSH) is the federal agency responsible for conducting research and making recommendations for the prevention of work-related injury and illness. NIOSH is part of the Centers for Disease Control and Prevention (CDC) in the Department of Health and Human Services.

NIOSH is the leading federal agency conducting research and providing guidance on the occupational safety and health implications of nanotechnology. This research focuses NIOSH's scientific expertise on answering questions that are essential to understanding safe approaches to nanotechnology. Some of these questions include:

- Are workers exposed to nanomaterials in the manufacture and use of nanomaterials, and if so what are the characteristics and levels of exposures?
- What effects might nanoparticles have on the body's systems?
- Are there potential adverse health effects of working with nanomaterials?
- What work practices, personal protective equipment, and engineering controls are available, and
- How effective are they for controlling exposures to nanomaterials?

NIOSH believes that the answer to these questions is critical for supporting the responsible development of nanotechnology and for maintaining competitiveness of the United States in the growing and dynamic nanotechnology market.

NIOSH publishes a document called *Approaches to Safe Nanotechnology*. The document reviews what is currently known about nanoparticle toxicity and control. The document serves as a request from NIOSH to occupational safety and health practitioners, researchers, product innovators and manufacturers, employers, workers, interest group members, and the general public to exchange information that will ensure that no worker suffers material impairment of safety or health as nanotechnology develops.

Opportunities to provide feedback and information are available throughout this document. You can download a copy of the document by contacting the following Web site: http://www.cdc.gov/niosh/topics/nanotech/safenano/

To learn more about Nanotechnology and NIOSH go to: http://www.cdc.gov/niosh/topics/nanotech/

NIOSH provides an online library on nanotechnology. Go to: http://www.cdc.gov/niosh/topics/nanotech/nil.html

NANO INTERVIEW: LAURA BLASI, Ph.D., AND NASA'S VIRTUAL LAB

Dr. Laura Blasi has been committed to the pursuit of equity in education, focused on evaluation in K-12 education and at the college-level. Her emphasis has been on improving innovation—such as the use of technology for teaching and learning. She has collaborated with colleagues at the Association of American Colleges and Universities (AAC&U) and at the University of Virginia's Center for Technology and Teacher Education (CTTE). In collaboration with her colleague Dr. Dan Britt, she is currently working on "To Infinity and Beyond" which emphasizes the history and role of telescopes and microscopes in the exploration of space, funded by NASA's Space Telescope Science Institute (STScI).

Dr. Blasi is a member of Project Kaleidoscope: Faculty for the 21st Century (F21), a national network of emerging leaders in undergraduate STEM (Science, Technology, Engineering, and Mathematics) focused on transforming the environment for learning for undergraduate students in mathematics and the various fields of science. Dr. Laura Blasi is currently an evaluation specialist at Saint Leo University near Tampa, Florida, working with faculty members in science to integrate NASA's Virtual Lab at the college-level and with faculty members in math to transform entry-level courses to support student success. While at the University of Central Florida (UCF) for the past 4 years, Dr. Laura Blasi's research had focused on high school science in low-income classrooms, specifically using the Virtual Lab simulation developed by NASA Kennedy Space Center.

You have been working with NASA's Virtual Lab, tell us more about this . . .

The Virtual Lab is a suite of simulations of advanced microscopes, including the Scanning Electron Microscope (SEM); Fluorescence Light Microscope; Energy Dispersive Spectrometer (for the SEM); and an Atomic Force Microscope (AFM). There are a range of specimens accessible through these tools, such as the eye of a housefly, euglena, an integrated circuit, and lunar dust. The Virtual Lab was developed by NASA within the Learning Technologies Project (LTP), and the software was designed and built by the Beckman Institute.

The software was refined using the findings from a study I conducted funded by the BellSouth Foundation in low socioeconomic schools using the Virtual Lab. The study documented the use of the Virtual Lab software within tenth grade science classrooms in the 2004–2005 school year (n=225) in low socioeconomic status areas in Orange County, Florida, while contributing to the further development of the program.

Working with my colleague, Theresa Martinez, at NASA Kennedy Space Center (KSC), we recently launched Cogs (Connecting a Generation to Science) on the Web (http://www.nasa-inspired.org/). This project seeks to develop materials to support the systemic integration of NASA's Virtual Lab into middle and high school classrooms nationally.

We developed Cogs from requests for support made by teachers when we piloted the Virtual Lab. While Cogs includes the Virtual Lab download and teaching materials to support the use of this simulation, teachers can post requests for new specimens to be added or they can ask questions that can be addressed by experts in the field.

How did the Virtual Lab develop?

A team from NASA Kennedy Space Center (KSC) led by Berta Alfonso proposed this Virtual Lab. The Virtual Lab concept was proposed in response to educators' requests to provide students access to NASA's instruments. Many of these instruments are too costly for schools and colleges to purchase and maintain. Specifically, this need was voiced within our discussions with Historically Black Colleges and Universities (HBCUs) in the United States.

By providing simulations of these microscopes, we could meet these needs in a way that did not conflict with KSC operations and security. By providing this software freely to all, NASA could also ensure equal access and availability to the local community, the nation, and beyond borders across the globe. Theresa Martinez now heads up the NASA KSC overseeing the Virtual Lab. We described the development of the Virtual Lab at length in the journal *Simulation and Gaming* last year (2006).

How did you get involved with the Virtual Lab?

I became involved as an independent evaluator. With funding from the BellSouth Foundation working with talented graduate students in the UCF College of Education, I conducted a usability study in three Orange County public schools that serve low-income communities using the Virtual Lab software in Florida. The schools were low performers on the state standardized tests for 3 or more years prior. So we were not focused on Advanced Placement students in mathematics and the sciences at well-funded schools for public relations purposes—we really wanted to know what was happening and if the technology could be improved to meet the needs of the students and teachers.

The Virtual Lab allows students and teachers to work individually with very sophisticated scientific instruments. These tools from NASA are free of cost and cannot be broken. They allow science teachers to teach science using a hands-on approach rather than only through lecture or textbook reading. This is the reason it was important for me to work within low-income schools, which provide access to the classroom conditions that are experienced across the country. If the limited access to technology is not addressed systemically in education, it will be difficult to reach the goals for the development and use of advanced technologies that we imagine setting and achieving over the next 10 years in the United States.

Can teachers become involved with the Virtual Lab project?
Yes, while teachers use the simulation in their classrooms, we also have had teachers then go on to train others in their schools. The Virtual Lab has also been developed for use in professional development for teachers in science and math, with the idea that teachers will more likely incorporate it if they have experience using and integrating the Virtual Lab into their lesson plans.

The interface includes annotation and measuring tools (from millimeters to microns) and the integration of the software can support student preparation for standardized tests in mathematics and the sciences. The Virtual Lab is also used at the college level in the sciences, it is not limited in its design but instead it replicates the actual microscope interfaces.

With this context in mind we have been developing Cogs, so that high school and middle school teachers can browse, print, and even create resources and lesson plans related to the Virtual Lab. The Cogs project also gives teachers access to related animations and videos on the tools themselves, their use, and careers in the field.

What are the career paths available for Science, Technology, Engineering, and Mathematics (STEM) education that emerged in the development of the Virtual Lab?
There are a few that may not be thought of as options for middle and high school students, simply because they are evolving with this generation and were not featured in the options that many teachers had available to them. Artists and programmers who created the Virtual Lab provide examples of emerging careers in design and development. Scientists and engineers worked as subject matter experts, providing information and insights—as well as virtual specimens for our project.

There are other roles for evaluators, teachers, and project directors directly related to STEM education, and these are roles that I have had throughout this project. It's important to keep in mind that both women and men are pushing the frontiers of this work. Career counseling and advising in high school and in college will play a large role when working toward these kinds of STEM careers, but earlier exploration paired with guidance is better to develop the necessary skills for success. In terms of nanotechnology, while aiming for specific goals in terms of the economic competitiveness of the United States, we also need to focus on ways students today can be part of the effort to improve the pathways toward those goals.

While the Virtual Lab can be used in middle, high school, and college classrooms, can programmers, scientists, and other researchers also get involved?

Yes, definitely. The Cogs Web site offers a place for microscopists to view requested specimens and to discuss ways of contributing to the project. Dr. Glenn Fried and Dr. Ben Grosser and their colleagues at the Beckman Institute who developed the software also have made sure that the architecture is open source on SourceForge. Defined in XML code, the Virtual Lab can be easily expanded by the developer community (see: http://virtual.itg.uiuc.edu/software/).

From your perspective, what role can the Virtual Lab play as we seek to prepare students in the United States to fuel innovation in the field of nanotechnology research and development?

The Virtual Lab specifically allows students to see objects that are familiar and relevant to everyday living at the level of millimeters and microns. In order to think and work at the level of nanometers, students need to understand the concept of scale while also experiencing the level of detail possible through advanced microscopy.

More generally in terms of careers in STEM, both men and women need to approach study as an active experience that they can do rather than as a secondary report that they can read about. The Virtual Lab allows students to experience first-hand investigation guided by their teachers using equipment that was once out-of-reach for almost all schools in terms of space available for equipment and expense.

There will continue to be debate over the ways to best prepare students for STEM careers, just as there will be debate over the role of virtual labs in education. We need to be clear that in both cases there are conclusive research findings and we are able to act on them. Overall, the Virtual Lab allows more diverse populations access to advanced microscopes across all socioeconomic status (SES) levels when adequate technology access and support is available.

NANO ACTIVITY: NASA VIRTUAL LAB. TRY IT!

Dr. Blasi explained in her interview that NASA provides a virtual lab. The virtual lab is a suite of microscopes and multidimensional, high-resolution image datasets, all available online, without cost for science classrooms, teachers, and students. The Virtual Lab can be used on computers without the purchase of any additional equipment. To download, A Description of the Virtual Lab Here: Go to: http://www.nasa-inspired.org/cogs/?q=nasavirtual

Animation of Atomic Force Microscopy Basics

What is an Atomic Force Microscope (AFM)? How does it "see" atomic resolution? How does it provide the Virtual Microscope with three-dimensional data about topography? This animation will show you the basics behind this imaging technique, and covers scanning tunneling, contact mode, and tapping mode AFM.

Download the Atomic Force Microscope Training Animation http://virtual.itg.uiuc.edu/training/#afm
For the Other Animated Tutorials Go to: http://www.nasa-inspired.org/cogs/?q=nasamicroscope

READING MATERIAL

Foster, Lynn E. *Nanotechnology: Science, Innovation, and Opportunity.* Upper Saddle River, NJ.:Prentice Hall, 2005.

Hall, J. Storrs. *Nanofuture: What's Next For Nanotechnology?* Amherst, NY: Prometheus Books, 2005.

Ratner, Daniel, and Mark A. Ratner. *Nanotechnology and Homeland Security: New Weapons for New Wars.* Upper Saddle River, NJ: Prentice Hall, 2003.

Scientific American (authors). *Key Technologies for the 21st Century: Scientific American: A Special Issue.* New York: W.H. Freeman & Co, 1996.

Shelley, Toby. *Nanotechnology: New Promises, New Dangers (Global Issues).* London, UK: Zed Books, 2006.

VIDEOS

Detecting Deadly Chemicals. Science Daily. Investigators on a crime scene can now use a new tool for collecting chemical or biological samples. The sampler gun collects samples on a cotton pad—eliminating direct contact with anything harmful, as well as risk of contaminating evidence—a GPS system to record the samples' location, a camera that snaps pictures for evidence, and a digital voice recorder and writing pad for taking notes. http://www.sciencedaily.com/videos/2006-12-10/

NASA Space Elevator. Can we build a 22,000-mile-high cable to transport cargo and people into space? http://www.pbs.org/wgbh/nova/sciencenow/3401/02.html

A Nanotechnology Super Soldier Suit. Nanotechnology in the military. http://www.youtube.com/watch?v=-pNbF29l9Zg&mode=related&search=

Detecting Deadly Chemicals. Science Daily. The Anthrax Scare and a new tool for collecting chemical or biological samples. http://www.sciencedaily.com/videos/2006-12-10/

Space Elevator. Scientists envision a space elevator based in the Pacific Ocean and rising to a satellite in geosynchronous orbit. http://www.sciencentral.com/articles/view.php3?article_id=218392162&language=english

AUDIO

Voyage of the Nano-Surgeons. NASA-funded scientists are crafting microscopic vessels that can venture into the human body and repair problems—one cell at a time. http://science.nasa.gov/headlines/y2002/15jan_nano.htm

WEB SITES

National Science Foundation: http://www.nsf.gov/crssprgm/nano/

NSF Home Page: http://www.nsf.gov

NSF News: http://www.nsf.gov/news/

Department of Homeland Security: http://www.dhs.gov/index.shtm

Department of Defense: http://www.defenselink.mil/

The National Institute for Occupational Safety and Health: http://www.cdc.gov/niosh/homepage.html

National Aeronautics and Space Administration: http://www.nasa.gov/home/index.html

Los Alamos National Laboratory: http://www.lanl.gov/

National Nanotechnology Initiative: The National Nanotechnology Initiative (NNI) is a federal R&D program established to coordinate the multiagency efforts in nanoscale science, engineering, and technology. The goals of the NNI are to: maintain a world-class research and development program aimed at realizing the full potential of nanotechnology; facilitate transfer of new technologies into products for economic growth, jobs, and other public benefit; develop educational resources, a skilled workforce, and the supporting infrastructure and tools to advance nanotechnology. http://www.nano.gov/

U.S. Food and Drug Administration: The U.S. Food and Drug Administration regulates a wide range of products, including foods, cosmetics, drugs, devices, and veterinary products, some of which may utilize nanotechnology or contain nanomaterials. http://www.fda.gov/nanotechnology/

National Institute of Standards and Technology: http://www.nist.gov/public_affairs/nanotech.htm

SOMETHING TO DO

You may enjoy a National Science Foundation game, called NanoMission. NanoMission™ is a cutting edge gaming experience which educates players about basic concepts in nanoscience. http://www.nanomission.org/

10

Colleges and Schools and Nanotechnology

In Chapter 1, Dr. Mihail Roco, a senior adviser for nanotechnology at the NSF's National Nanotechnology, said that companies building products at the atomic level eventually would face a serious shortage of talent—far worse than what is already occurring.

Dr. Roco estimated about 2 million nanotech-trained workers will be needed to support growing industries and the startups they start within the next 10 to 15 years. He emphasized that the country needs to find ways to motivate students about sciences and to make them aware of the career opportunities in nanotechnology fields.

Based on the need to employ 2 million nanotechnology-savvy workers by 2015, the National Science Foundation (NSF) is pushing for children between the ages of 10 and 17 to become educated now about the field of nanotechnology. The science organization also stated that 20 percent of the 2015 workforce would be scientists. The remaining 80 percent will consist of highly skilled engineers, technicians, business leaders, mechanics, sales representatives, graphic designers, economists, and others. So, many workers will not need a Ph.D. to get into this field.

> Work Force Preparation. Are We Prepared to Get Into the Nanotechnology Workforce? Professor Wendy Crone. Go to: http://mrsec.wisc.edu/Edetc/cineplex/MMSD/prepared.html

To learn more about career paths in nanotechnology, you may want to watch the following video that interviews a number of students and professionals about their nanotechnology careers: http://virtual.itg.uiuc.edu/training/index.shtml#careers

This chapter will highlight some middle schools and high schools, several universities and colleges, and a few nonprofit organizations on how they are helping the young people understand nanotechnology and to explore careers in this emerging field.

AN INTERVIEW WITH DR. NANCY HEALY, EDUCATION COORDINATOR OF THE NATIONAL NANOTECHNOLOGY INFRASTRUCTURE NETWORK (NNIN)

One of the major groups that is providing state-of-the-art nanotechnology facilities and resources to schools and colleges is the National Nanotechnology Infrastructure Network (NNIN). The author had an opportunity to talk with Dr. Healy during a phone interview about her professional career as the Education Coordinator of the National Nanotechnology Infrastructure Network (NNIN). Dr. Healy's office is located at Georgia Institute of Technology, Atlanta, Georgia.

The NNIN is a National Science Foundation funded program, which supports nanoscience researchers by providing state-of-the-art nanotechnology facilities, support, and resources. The NNIN is a consortium of thirteen universities across the United States (http://www.nnin.org). In addition to researcher support, the NNIN has a large and integrated education and outreach program.

What is your role as the Education Coordinator of the National Nanotechnology Infrastructure Network?

The NNIN education programs focus on the development of a nano-ready workforce as well as development of a nano-literate public. Outreach efforts span elementary-level students through adults. The NNIN provides a variety of programs which include: summer camps for middle and high school students; on-site and off-site school visits which include laboratory tours, hands-on activities, demonstrations, and presentations; summer research experiences for undergraduates and K-12 teachers; workshops for K-12 teachers; K-12 instructional materials development; workshops/seminars for undergraduates; community college programs; symposia at national meetings; workshops for faculty, industry, and government personnel (lifelong learning); and a Web site for accessing information on our resources and programs.

Where did you grow up and what were your favorite subjects in high school?

I grew up in Greenville, Rhode Island, and my favorite subjects in high school were science and history. I wanted to be an archaeologist so I could combine the two subjects.

What college did you attend and what was your major?
I attended University of Rhode Island for my undergraduate degree in zoology and then I received a master's degree and Ph.D. degree in geology at the University of South Carolina.

How did you get interested in a career in nanotechnology education?
I don't think anyone working in nanotechnology education had a straight path in this career. I knew about nanotechnology when I was in South Carolina working for the state agency that oversaw all the colleges and universities. We had just approved the Nano Center at the University of South Carolina. So I learned about the field of nanotechnology. Around the same time, I found out that Georgia Tech wanted somebody to coordinate this new program in nanotechnology education and outreach (NNIN). They wanted someone with an interdisciplinary science background, which is what I had, and someone who had an understanding of teacher professional development in K-12 math and science. I applied and I received my new position in July 2004.

You were quoted in an article, "People generally don't know what nanotechnology really is. There's a risk that their perceptions will be based on popular culture portrayals of it rather than fact." What are some ways to help the general population to be able to separate nanotechnology fact from fiction?
I really think it is the responsibility of researchers, whether they are in the university, in government, or in industry to make sure that the public knows what is true and what is not true about any field, whether it is nanotechnology, biomedicine, or aeronautics. Researchers and scientists need to engage the public about what they are doing. One way of communicating this to the public is to have scientists address various clubs and organizations such as the rotary clubs. Having nanotechnology forums in schools and campuses is another way to educate the public. I think it is very important to work with the media, too. Scientists have the responsibility to make sure their science reported in the media is accurate and correct, and to correct any misconceptions in their work.

What are some of the benefits in nanotechnology? What are some of the ethical or societal issues?
One major benefit will be in the microelectronic and telecommunication fields. Our whole electronic world will be faster and smaller and will consume much less energy than we do now. Another benefit will be in the area of nanomedicine. This is a huge area. We will see some amazing

things happening in biomedical applications, particularly in such areas as cancer detection and treatment.

Dr. Nancy Healy, Education Coordinator of the National Nanotechnology Infrastructure Network (NNIN). The NNIN is a National Science Foundation funded program that supports nanoscience researchers by providing state-of-the-art nanotechnology facilities, support, and resources. (*Courtesy National Nanotechnology Infrastructure Network*)

On social and ethical issues, I think one of the major concerns is about the production of materials using nanoparticles. The question is, "will the nanoparticles be the next asbestos problem?" There is research that says yes they are harmful, but I have spoken with other researchers who point out that some of these studies are not done well. So it is very important that good studies and good science be done in this area of concern. No one wants to breathe in bad stuff into our lungs, but everyday we intake particles from vehicle exhausts. So will the Buckyballs be any more harmful than the soot that comes out of the exhausts from vehicles? That is a study that needs to be done. The EPA and NSF are funding more and more research programs to look at the safety concerns of nanotechnology.

I think that there is a big ethical issue of nanotechnology that could occur between the haves and have-nots. In other words, will the industrialized nations get all this technology, and in doing so, bypass those countries that are poorer? This issue needs to be addressed by everyone in the industry and that includes the scientists, ethicists, and the lawyers.

In the past the NNIN has presented programs at the National Science Teachers Association to inform teachers about nanotechnology. What kind of presentation have you done?

In NNIN, we have a program called the NNIN Research Experience for Teachers (NNIN RET). The teachers who were participants in this program have presented instructional nanotechnology units at the

NSTA meetings. The teachers developed these units as a result of their summer nanotechnology research experiences at such colleges as Georgia Tech, Harvard, Howard University, Penn State, and the University of California, Santa Barbara. In 2006, 19 teachers participated in the NNIN RET and their units were presented to schoolteachers at the 2007 NSTA. After the presentations, the schoolteachers visited our NSTA exhibit booth to review the units and to get additional information. These units are posted or will be posted on our Web site.

One teacher, from Gwinnett County, Georgia, created a physics and literature unit based on material in the Ratner and Ratner book, *Nanotechnology, A Gentle Introduction to the Next Big Idea.* Even though she teaches advanced placement physics, she included literature materials into the unit. Then, she brought the unit to the elementary grades by having her students create books on what the future will be like with nanotechnology. In the beginning, the NNIN RET involved all high school teachers with physics and chemistry backgrounds. We now have added middle school teachers with backgrounds in physical science and in basic biology.

The NNIN program sponsors volunteers who go to local schools and work with student groups who visit the Georgia Tech Campus. What kinds of activity lessons do they present? Are any of these lessons available for teachers?

Yes, lessons are available to teachers. As an example, we have our "Modeling Self-Assembly" lesson and another lesson on hydrophobic and hydrophilic properties posted on the education portal. We are now preparing and working on another 20 units that will be available to teachers.

We also have a teacher resource flyer that we hand out to teachers and that flyer is available on the Georgia Tech Web site (http://www.mirc.gatech.edu/education.php). The NNIN goal is to provide information to teachers so they can go to one place for specific information on nanotechnology units, such as a chemistry unit, that they can use in their classes.

The NNIN program has a variety of summer camps for high-school aged students. What do the students learn at these camps?

Every camp we have provides an introduction to nanotechnology. We give the kids an understanding of what nano is and where it is going. Every camp has information on education and career opportunities. So, kids get an idea what the field is, where it is going, and what jobs will be available. All the camps bring the kids into some kind of laboratory facility. At Georgia Tech, they visit at least three research laboratories learning about what the researchers are doing in their own labs. Students

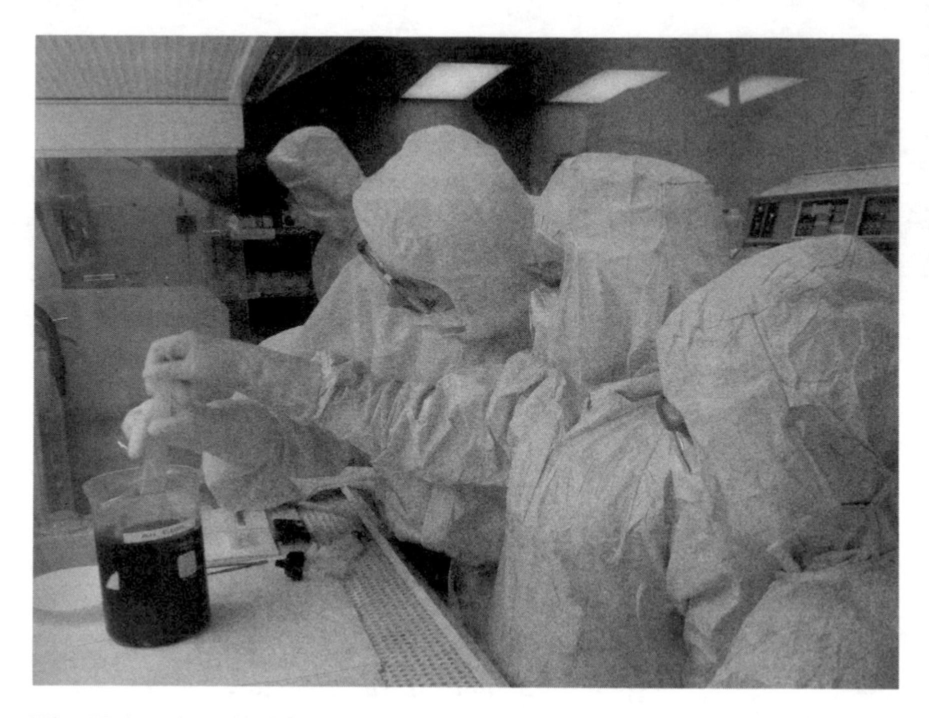

The University of Michigan conducts a nanoelectronics camp for students. In this photo, students are working with a professor in a cleanroom environment. (*Courtesy National Nanotechnology Infrastructure Network*)

also visit a cleanroom facility to learn how it is used in nanotechnology. At some camps, students can do microchip preparation using a cleanroom. The camps provide a lot of exposure in the nanotechnology field.

Do students at these camps have access to different kinds of electron microscopes?

Yes, at some camps. We are moving into this area of developing ways for students to access Atomic Force Microscopes and Scanning Electron Microscopes. Some groups at the University of California, Santa Barbara, have been able to remotely use the AFM at Penn State.

We had a group of high school students come in and use our AFM and our SEM microscopes at Georgia Tech. These microscopes allowed the students to go down and look at miniscule amounts of matter. We are also looking at ways to use remote access to these microscopes, too.

Penn State has an animation of how a scanning electron microscope is used. This animation, *Amazing Creatures with Nanoscale Features,* is an introduction to microscopy, scale, and applications of nanoscale properties. It introduces some of the tools that are used by scientists to visualize samples that are smaller than what we can see with our eyes. This activity is available for use via the Center's Web site at http://www.cneu.psu.edu/edToolsActivities.html.

I learned that one college had a nanotechnology career day and it attracted more than 300 school-age students. The NNIN also has career days. What are some of the advantages for students to attend these kinds of career day events?

We do something called *Nanoinfusion* at Georgia Tech. This event is pretty much like a career day. At one of these events, we had about 375 eighth grade students split between 2 days for about 4 hours each day.

The activities of a nano career day, or field trip to a facility, depends on which facility they visit and the duration of the visit. Typically there are hands-on activities and demonstrations, lectures, and tours. The advantage of all these nano career days is that it gets kids on a college campus, and for many, to see a college campus for the first time. While on campus, the students can interact with graduate students and faculty members. To excite students, the career day should have a tie-in with something that has real-world applications which is what we strive to provide our visitors.

Diana Palma, the Assistant NNIN Education Coordinator at Georgia Tech, is the one who developed and organized *Nanoinfusion*. She can keep 150–175 kids busy during a 3–4 hour visit. The students visit our facility and they go to a series of stations that are set up where they get to see either a demonstration or do a hands-on activity at each station. Then they move on in groups to experience different things, such as a cleanroom tour or a laboratory tour. For the first *Nanoinfusion*, we had 15 research labs on campus participate and 70 volunteers to assist the students. At the end, we may have them create an advertising piece or commercial based on a nanotechnology product that they have seen or a possible product from someone's research. They get exposed to a whole range of things during their visit.

NNIN has featured on its Web site a number of features that are directed to students and teachers. A partial list includes Nanotechnology Careers, Nanotechnology Tools, Nanotechnology Products, Seeing Nanostructures and even a children's science magazine called *Nanooze*. Is the NNIN planning to add any additional educational topics?

We are planning on a few more features and to provide additional nanotechnology educational units on the NNIN Web site for K-12 teachers.

What would your advice be for high school students who wish to explore a career in nanotechnology?
At the high school level, students need to take a minimum of 3 years of science and mathematics. They need to explore some of the career and educational options that are available in this field. I wish more states would duplicate the program at Penn State. Penn State has a special program where students can obtain a two-year degree in nanotechnology (at a local community college with a capstone semester at Penn State) and then go on to a 4-year program if they want to pursue further education. Students should know that they do not need to have a Ph.D. degree to work in the field of nanotechnology research. As with any field, there are all kinds of opportunities. At the technical level, they could be maintaining vacuum pumps in a lab or learning how to do different fabrication processes in the cleanroom. You can be a lawyer and be involved with nanotechnology perhaps as a patent lawyer. You can be a graphic artist or a businessperson on the entrepreneurial side. But you need to have some basic understanding of the interactions of all the sciences and engineering. One of the units we have up on our Web site, which is our nanoproducts unit, includes an extension activity on careers. The unit provides the students with a number of sites they can visit to explore career opportunities. We eventually plan to have on our Web site a whole unit on exploring nanotechnology careers.

You can visit the NNIN Web site at: http://www.nnin.org

NATIONAL SCIENCE TEACHERS ASSOCIATION AND EXPLORAVISION

Each year, several elementary, middle school, and high schools have entered science projects in a program called ExploraVision, which is sponsored by the Toshiba Corporation and the National Science Teachers Association (NSTA). The ExploraVision is designed for K–12 students of all interest, skill, and ability levels. ExploraVision encourages students to create and explore a vision of future technology by combining their imaginations with the tools of science.

Since 1992, more than 200,000 students from across the United States and Canada have competed in ExploraVision. For many of the participants ExploraVision can be the beginning of a lifelong adventure in science, as students develop higher-order thinking skills and learn to think about their role in the future.

Each year there are several schools that explore a future vision of nanotechnology. As an example, in Chapter 5, the author interviewed students who were winners of the ExploraVision Program. There have been other ExploraVision nanotechnology winners over the past years. Here is a short list of some of those schools that participated in the ExploraVision program.

ExploraVision Project

The Tumor-nator, Mr. Paul Octavio Boubion. Mr. Paul Octavio Boubion is an eighth-grade physical science teacher at the Carl H. Lorbeer Middle School, in Diamond Bar, California. I asked Mr. Boubion to comment on his experiences in the nanotechnology field and the ExploraVision program. Here is what he had to say.

> I got interested nanotechnology from my brother-in-law, Timothy Ryan, PhD. who is a research scientist at Cornell University. Dr. Ryan is a neurophysicist and has inspired me with his research and discoveries. He is a wealth of information about a wide variety of scientific topics and my "lifeline" when I have a question.
>
> I have also been inspired by the creative topics my students have researched for Toshiba/NSTA ExploraVision contest each year. The best entries my students have submitted in the past seem to be nanotech related.
>
> The "Tumor-nator" was a Toshiba/NSTA ExploraVision project and was one of the main inspirations as a teacher. The four 8th graders worked on this project for months and I provided as much help as I could. The students learned more about science through their research than all they learned in middle school.
>
> The students project, Tumor-nator, received a Toshiba/NSTA ExploraVision Honorable Mention in 2005. The tumor-nator targets malignant tumors in the body and repairs the P53 gene in the DNA of cancer cells. The P53 gene triggers cell death in the life cycle of a healthy cell. Then, T-cells can work to remove the cells using the body's own natural processes. The Tumor-nator treatment also contains, interferon, which stimulates T-cell production in the body.
>
> My advice for students who want to explore a career in nanotechnology is to take their laboratory training very seriously so they are prepared to do science, not merely earn a grade.

You can contact Mr. Boubion at this email: paul.boubion@pusd.org

ExploraVision Project: Appy-Bot, Norma L. Gentner

Mrs. Norma L. Gentner is an Enrichment Teacher (K-5) at Heritage Heights Elementary, Sweet Home Central School District, Amherst, New York. Mrs. Gentner's team of fifth-grade students took second place in the Nationals as ExploraVision winners. I asked Mrs. Gentner to comment on her experiences in the nanotechnology field and about

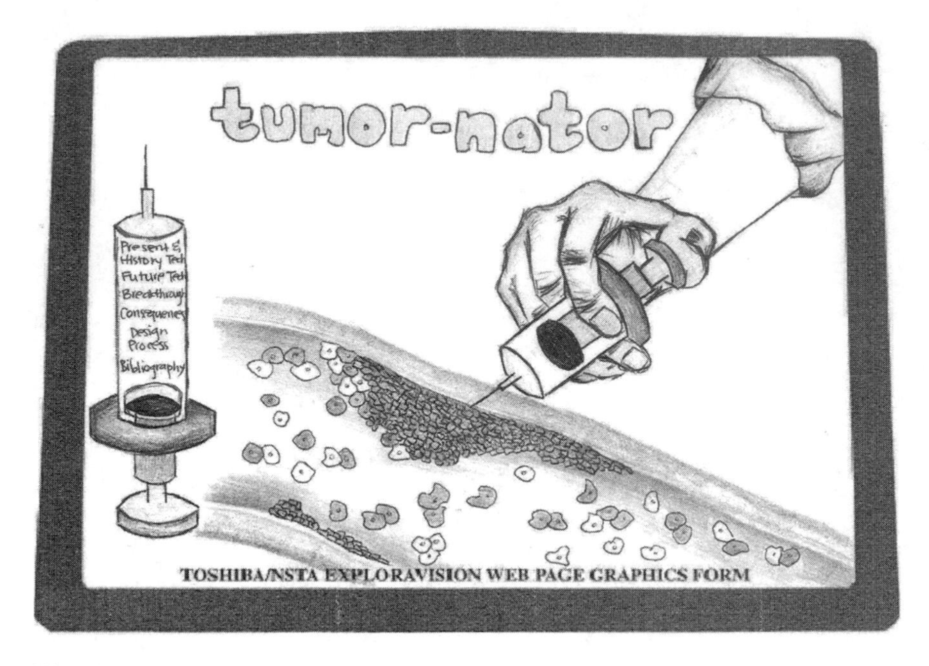

The "Tumor-nator" was a Toshiba/NSTA ExploraVision project done by eighth-grade students in Diamond Bar, California. Their teacher is Paul Boubion. The Tumor-nator targets malignant tumors in the body and repairs the P53 gene in the DNA of cancer cells. (*Courtesy Paul Boubion*)

the award-winning ExploraVision project, the Appy-Bot. The following includes her comments,

> The Appy-Bot will eliminate the need for appendectomies in the future by using nanotechnology to repair a patient's appendix.
>
> Normally, nanotechnology wouldn't be a science studied at the elementary school level, but the national science competition, ExploraVision provides a forum that challenges students and teachers to go beyond the classroom curriculum.
>
> During the 2004–2005 school year, a team of three fifth graders wanted to create a future technology that would eliminate the need for appendectomies. The Pediatric Surgeon who performed an appendectomy on one of the team members came in and shared what was happening to the appendix before surgery. He also encouraged the team to do some research on nanobots. They went on-line and learned about nanobots and nanotechnology.
>
> The next part of the design process took them to the University of Buffalo where they met with professors in Chemical and Biological Engineering. One professor felt that the team's initial concept for the Appy-Bot had too many functions for it to be nano-sized and encouraged them to create micro-machines. The competition required them to look 20 years into the future,

Norma L. Gentner with her team of fifth grade students who were Second Place ExploraVision winners. Their nanotechnology project was called, The Appy-Bot, which eliminates the need for appendectomies in the future by using nanotechnology to repair a patient's appendix. (*Courtesy Norma L. Gentner*)

and they felt that micro-machines were not futuristic enough. They also discovered that the 2020s were going to be the "Golden Age of Nanobots," which would be the perfect time to launch their idea. The team left that professor's office disheartened, but ran into another professor from that department Mrs. Gentner had used before. He took them to a conference room and "really" helped them to understand through analogies and some math calculations just how small a nano was. Their project went through its final revision, and became a fleet of millions to billions of nanobots in a gel cap someone would swallow. There would be two types of Appy-Bots. One type would suck in surrounding liquids, flushing out the obstruction to the appendix, while the other would pick away at it.

The team took 2nd Place in the Nationals. National Winners and their Web sites/projects are posted on the ExploraVision Web site (http://www.exploravision.org). Or, you can go directly to the team's Web site (http://www.exploravision.org/past-winners/winners-2005.php). Just scroll down the page until you get to "The Appy-Bot," and click on "View winning Web site."

Again, you asked how important nanotechnology is in the subject I teach. As you can see, it was extremely rewarding in its application to ExploraVision challenges, and future science. Due to the abstract nature of nano science though, not all students at the elementary level would be as ready as this team was in its application. Plus, most of the science taught at this level concerns itself with present science, and doesn't usually take a problem-solving approach. This was real-life application!

I think that the possibilities of what nanotechnology can do for medicine in the future and people in general are truly exciting. But, we have to take a different approach to science in elementary schools, where students are applying what they learn in the classroom to authentic questions/problems.

You can go to their Web site at http://dev.nsta.org/evwebs/531/

Other Toshiba/NSTA ExploraVision Winners

La Jolla, California

The La Jolla High School in La Jolla, California, developed *The Nanoclotterator*. The nanoclotterator is an injectable blood-clot eating nanobot designed like the T4 bacteriophage virus. The NCR would be about 4 microns in size—small enough to fit within the tiniest capillaries, yet large enough to keep from passing through blood vessel walls. A specially designed "mouth" would extend from the NCR capsule, biting out and removing chunks of a blood clot. The school was a second place ExploraVision winner in 2006. The project Web site is http://dev.nsta.org/evwebs/3314/

St. Louis, Missouri

Middle School students at the John Burroughs School, St. Louis, Missouri, developed a way to help travelers overcome airline jet lag and students can study all night. The name of their project was called the Body Clock Band. A nanocomputer imbedded in the wristwatch-like device dispenses sleep-inducing or "stayawake" pharmaceuticals through the skin according to user-programmed sleep cycles. Drugs are nonhabit forming and safeguard excessive use.

You can go to the ExploraVision Web site to find other Toshiba/NSTA winners in a variety of fields besides nanotechnology.

Did you know?
The Toshiba Corporation, based in Tokyo, is a diversified manufacturer and marketer of advanced electronic and electrical products, communications equipment, power systems, industrial and social infrastructure systems, and household appliances.

COLLEGES MOTIVATE STUDENTS IN NANOTECHNOLOGY

Many universities and colleges sponsor a variety of workshops, career days, field trips, laboratory and cleanroom visits, hands-on nanotechnology activities, and Web sites to motivate students in science and nanotechnology.

NanoDay at Northwestern

The Northwestern University-Nanoscale Science and Engineering Center sponsors a NanoDay at their campus. The mission of the NWU-NSEC is to foster a lifelong interest in science and technology by teaching people of all ages about the nanoworld. Many schools have come to Northwestern to participate in "NanoDay." It is a half-day of activities designed to spark student interest in nanoscience and technology. The students watch a variety of demonstrations led by professors. The students also participated in a hands-on activity: building model LEGO® atomic force microscopes (AFMs). At the end of the activities, each student received a take-home "Try This!" Nanokit developed by the University of Wisconsin-Madison. For more information contact: www.nsec.northwestern.edu

Northwestern University and Materials World Modules

Materials World Modules is a National Science Foundation (NSF) funded inquiry-based science and technology educational program based at Northwestern University.

The program is dedicated to helping teachers enhance their science curriculum and excite their students about science and the world we live in. The Materials World Modules (MWM) Program has produced a series of interdisciplinary modules based on topics in materials science, including Composites, Ceramics, Concrete, Biosensors, Biodegradable Materials, Smart Sensors, Polymers, Food Packaging, and Sports Materials. The modules are designed for use in middle and high school science, technology, and math classes. Over 9,000 students in schools nationwide have also used them. For more information, go to Web site: http://www.materialsworldmodules.org/aboutmwm.htm

You may be interested in seeing a video: Inside MWM: An Interview with Lake Forest Science Teacher Kate Heroux. In this interview with science teacher Kate Heroux, Professor Robert Chang, the developer of MWM, discusses the successes and challenges of implementing the Materials World Modules in high school. To see the video go to: http://www.materialsworldmodules.org/videocenter/MWMinterview_kh.htm

Penn State Nanotech Camps

Penn State offers programs for high school students interested in learning more about nanofabrication. These include one-day experiences during the school year, and the extremely popular, three-day

Nanotech Camps during the summer months. Throughout the year, students in grade 7 and above are invited to attend 1-day camps at University Park to receive an introduction to nanotechnology. For students in ninth grade and above, three-day summer nano camps offer more in-depth opportunities for looking into careers, education, and the details of nanotechnology fabrication and application. Penn State offers students certificates or degrees in Nanofabrication Manufacturing Technology (NMT). You can also request a free copy of Careers in Nanotechnology Information Video at www.cneu.psu.edu

Rice University: The NanoKids™

The NanoKids™ is an educational outreach program, headed by Dr. James M. Tour, Chao Professor of Chemistry and Director of the Carbon Nanotechnology Laboratory at Rice University. The NanoKids program is dedicated to increasing public knowledge of the nanoscale world and the emerging molecular research and technology that is rapidly expanding internationally. Some of the goals in the program are to:

- significantly increase students' comprehension of chemistry, physics, biology, and materials science at the molecular level;
- provide teachers with conceptual tools to teach nanoscale science and emerging molecular technology; and to
- demonstrate that art and science can combine to facilitate learning for students with diverse learning styles and interests.

For more information about NanoKids go to: http://cohesion.rice.edu/naturalsciences/nanokids/

University of California, Santa Barbara: Chip Camp

The University of California offers a program called Nanotech Chip Camps for small groups of high school students. The Nanotech Chip Camp at UCSB pairs high school students with graduate student mentors in a cleanroom laboratory and teaches them nanofabrication processes. On the first day, students make a micro resistor. On the second day, students have a guided exploration of parameters and do an analysis. Students also attend talk(s) provided by faculty member(s) about various aspects of nanotechnology. Students will then be able to share what they have learned through this experience via their school or county science fair. By using scientific inquiry, students will have a better understanding of the intricacies of nanofabrication processes and how these processes relate to other pure and applied sciences, as well as the world around them.

The Chip Camp provides students (and the participating teacher) with an intensive two-day laboratory experience that enables students to:

- gain a better understanding of integrated circuit systems;
- understand how the order of steps used in processing affects the outcome;
- develop procedures which compare change with constancy;
- model the processes and explain it to someone else; and
- organize what they have learned into a comprehensive display.

For more information about Chip Camps write to berenstein@ ece.ucsb.edu

College of Nanoscale Science and Engineering (CNSE), University of Albany

The College of Nanoscale and Engineering (CNSE) has been assisting school districts in the implementation of programs that encourage science awareness in grades K-12 and help build the science and technology workforce of the future.

In October 2006, CNSE announced the $500,000 "Nano High" initiative program. The mission of this pilot program is between CNSE and the City School District of Albany. The goal is to develop and implement innovative science and engineering educational programs between Albany High School and CNSE. Under a pilot program, NanoHigh will focus on school-to-work activities designed to train Albany High School students in creative nanoscience and nanoengineering concepts. The program will also help equip them with the skills necessary to pursue advanced educational opportunities in the emerging field of nanotechnology. For more information go to: http://www.albanynanotech.org/

Cornell University

Cornell University sponsors a program called Main Street Science. Main Street Science serves the needs of the K-12 students and advances the science literacy of the general public through the development of hands-on science, technology, engineering, and math (STEM) activities. Main Street Science provides lesson plans on cantilevers, diatoms, microscopy and scale, motors, sunprint paper and photolithography, sodium alginate polymers, and chromatography. Go to: http://www.nbtc.cornell.edu/mainstreetscience/index2.html

The University of Wisconsin-Madison

The University of Wisconsin-Madison Materials Research Science and Engineering Center (MRSEC) Interdisciplinary Education Group (IEG): uses examples of nanotechnology and advanced materials to explore science and engineering concepts. The center provides K-12 curriculum kits and activities, programs, exhibits, and videos. They even have a slide-show library.

You can reach their Web site at: http://www.mrsec.wisc.edu/edetc/modules/index.html

Arizona State University

The Arizona State University sponsors a program called SPM - LIVE! The program gives educators and researchers the opportunity to perform Scanning Probe Microscopy (SPM) experiments, through the Internet with image broadcasting and control of the instrument on a real-time basis. They offer a list of modules that include size and scale, and scanning probe microscopy, the Allotropes of carbon and engineered materials. For more information send an email: invsee@asu.edu.

North Carolina State University and the University of North Carolina, Chapel Hill

The universities provide a unique software program where you can explore characteristics of viruses with the use of atomic force microscopy. Your challenge in the software program is to figure out what is making a puppy ill. You observe a phage invading a bacterium, learn about how viruses take over a cell and multiply, and examine different virus structures. The program also includes the history of virology research. This program is available free to educators and students for use in instructional contexts. For more information go to: http://ced.ncsu.edu/nanoscale/materials.htm

SRI International: NanoSense

As you learned in Chapter 6, The SRI International provides a program called NanoSense. Working closely with chemists, physicists, educators, and nanoscientists, the NanoSense team is creating, classroom testing, and disseminating a number of curriculum units to help high school students understand science concepts that account for nanoscale phenomena and integrate these concepts with core scientific ideas in traditional curricula.

NanoSense is led by Co-Principal Investigators, Patricia Schank and Tina Stanford at the Center for Technology in Learning. SRI International is an independent, nonprofit research institute conducting client-sponsored research and development for government agencies, commercial businesses, foundations, and other organizations.

For more information about Nanosense, or if you have questions or comments, contact us at: nanosense-contact @ ctl.sri.com

NANOTECHNOLOGY IN MUSEUMS AND TRAVELING EXHIBITS

Museums also are developing exhibits, demonstrations, and activities for people of all ages to learn about the field. The Boston Museum of Science offers programs, media, and exhibits about recent research in nanoscale science and nanomedicine. Both online and in the Exhibit Halls, the Museum is an information clearinghouse for this emerging field, utilizing podcasts, cablecasts, live demonstrations, and special events to bring visitors the latest nanotech information.

The museum offers live presentations that have included Nano-Sized Factories of the Future, Nanotech Today, and Nanoparticles for Cancer Drug Discovery. Contact: http://www.mos.org/topics/cst

Making Things Smaller

Purdue University's Department of Physics and the School of Electrical and Computer Engineering, in conjunction with the Children's Museum of Oak Ridge, offers an exhibit called "Nanotechnology: The Science of Making Things Smaller." In the exhibit, Youngsters will have the opportunity to learn firsthand about nanotechnology through video animations, a wall of nano-art, hands-on activities, posters, and a ***LEGO®*** scanning probe microscope. In fact, the first stop in the exhibit for many visitors is the ***LEGO®*** working model of a scanning probe microscope. The ***LEGO®*** model "sees" by probing objects with a small, sharp tip. By slowly touching the ***LEGO®*** model surface with a tip, a three-dimensional image can be generated and displayed on a computer.

It's a Nano World and Too Small to See Exhibits

Cornell University has a traveling nanotech exhibit called, It's a Nano World. Over the last 3 years, elementary school children all over the United States have been learning about incomprehensibly tiny things by walking through and playing with very large and colorful things in a traveling science museum exhibition. The exhibition was created by Cornell University's National Science Foundation (NSF)-funded Nanobiotechnology Center (NBTC) in partnership with the Sciencenter, Ithaca's

Sciencenter has a traveling nanotech exhibit called *It's a Nano World* and another one called *Too Small To See.* Attending the exhibits, school-aged children have been learning about tiny things by walking through and playing with very large and colorful objects in a traveling science museum exhibition. In this photo, a visitor at the exhibit is discovering that molecules are too small to see, but not too small to smell. (*Courtesy Gary Hodges and the Sciencenter, Ithaca, NY*)

hands-on science museum, and Painted Universe, a local design firm. The focus of the exhibit, aimed at 5- to 8-year-olds and their parents, is to explain concepts of size and scale, showing children that many things in the world are too small to be seen with the naked eye.

It's a Nano World exhibit first opened at the Sciencenter in 2003. Since then the exhibit has traveled to science museums in Ohio, South Carolina, Louisiana, Michigan, Virginia, and Texas. An estimated 3 million people have seen the exhibit.

Another exhibit developed by the same group at Cornell is called, "Too Small to See," This exhibit is aimed at middle school students to explain how nanotechnologists create and use devices on a molecular scale. The exhibit, aimed at 8- to 13-year-olds, helps visitors view the world at the atomic scale and to better understand just how small a nanometer is.

Renee DeWald is a chemistry teacher at the Evanston Township High School, Evanston IL. Her high school team researched and developed a project titled, Fighting Cancer with Nano-Shells, that in the future, nano-shells would be used as one of the major treatments for cancerous tumor cells. (*Courtesy Renee DeWald*)

Lawrence Hall of Science

Established in 1968, at the University of California, at Berkeley, in honor of Ernest O. Lawrence, the Lawrence Hall of Science provides an online site, called the Nanozon. At this site you can do a variety of activities such as solving nanopuzzles, watching a nanovideo, make a nanometer ruler, and even ask Nanobrain a question such as how tall is a cupcake or how small is a molecule. http://nanozone.org/

HIGH SCHOOL PROGRAMS IN NANOTECHNOLOGY

The Future is N.E.A.R. program at North Penn High School in Lansdale, Pennsylvania, offers its students an opportunity to gain 21st century skills that will help prepare them to become successful leaders in the new, technological global society. The program introduces the fundamentals of nanotechnology, engineering research, and higher-level thinking and application of knowledge to high school students while cultivating their interest in engineering, problem solving, and life-long learning. Some of the main goals of The Future is N.E.A.R. project are to inform and educate students about nanotechnology, the present and future applications of nanotechnology and to improve science, technology, engineering, and mathematical (STEM) knowledge and skills utilizing a collaborative, real-world research environment.

NANOTECHNOLOGY INFORMATION AND SITES FOR TEACHERS

The Nanoscale Informal Science Education (NISE) Network brings researchers and informal science educators together to inform the public about nanoscience and technology. The network includes an online resource center that features careers in nano and even activities designed for students who have an understanding of atoms and molecules to help them learn the applications of nanotechnology in everyday life.

The National Center for Learning and Teaching in Nanoscale Science and Engineering (NCLT) is the first national center for learning and teaching of nanoscale science and engineering education in the United States.

The mission of NCLT is to develop the next generation of leaders in NSE teaching and learning, with an emphasis on NSEE capacity building, providing a strong impact on national STEM education.

The NCLT provides Professional Development Opportunities for 7–12 grade science teachers interested in learning more about Nanoscience and how to incorporate this cutting-edge research into the classroom.

This program offers science teachers opportunities to work with nanoscience researchers, science educators, and learning tool experts in content development and laboratory experiments. For more information about this program: http://www.nclt.us/pd_activities.htm

With support from the National Science Foundation, the NanoScale Science Education Research Group has been investigating how students learn about nanotechnology and nanoscale science. This research includes studies of elementary, middle, high school, undergraduate, and graduate students. For more information about this program and a listing of studies go to: http://ced.ncsu.edu/nanoscale/materials.htm

NANO INTERVIEW: MS. RENEE DEWALD, TOSHIBA/NSTA EXPLORAVISION, FIGHTING CANCER WITH NANO-SHELLS

Renee DeWald is a chemistry teacher at the Evanston Township High School, Evanston, Illinois. The high school team researched and developed a project titled, Fighting Cancer with Nano-Shells, that in the future, nano-shells would be used as one of the major treatments for cancerous tumor cells. The ExploraVision project, Fighting Cancer with Nano-Shells, earned an honorable mention in 2006.

How did you get interested nanotechnology?
For many years, I have worked with Northwestern University Professor Robert Chang on the Materials Worlds Modules curriculum program. When NU became a partner in the first national center for learning and teaching of nanoscale science and engineering education in the United States (NCLT), I became interested in nanotechnology and facilitated summer professional development opportunities for high school teachers in nanoscience.

Describe some of your teaching experience in developing the nanotechnology project.
For many years, I have had students enter the Toshiba ExploraVision competition. Working in groups to simulate research and development teams, students select a current technology and after researching its history and present, creatively project what the technology will look like in 20 years. They predict what breakthroughs will be needed to realize the vision as well as examining the possible positive and negative consequences of the new technology.

Each year, several of my student teams would earn honorable mention awards, and all of these winning projects incorporated nanotechnology in the vision. So more recently, I started actively teaching a nanotechnology unit before the project and encouraged all students to use nanotechnology concepts in their ExploraVision papers. The project Fighting Cancer with Nano-Shells earned an honorable mention in 2006.

How important is the field of nanotechnology in the subject you teach?
I think nanotechnology has the potential of teaching students the how of science. Often we tell students what we know without telling them how we know. In the nanoscience unit that I teach, students build a model of an atomic force microscope after learning about atomic structure.

What advice would you give students who want to explore a career in nanotechnology?
I would tell them that nanotechnology is the future of science and that the worldwide need for nanotechnology workers is expected to reach 2 million by 2015.

Can you provide a Web site or recommend a one where our readers would be able to access more information about your project work?
Yes, I can. Here are a few of them.

www.exploravsion.org has information on the contest as well as Web sites of past winning papers. Here is another Web site:

http://www.nclt.us is the Web site for the NCLT and a valuable nanotechnology resource.

NANO ACTIVITY: NANOTECHNOLOGY CONSUMER PRODUCTS

The National Nanotechnology Infrastructure Network (NNIN) provides a Web site titled, Nanotechnology Educational Resources and Activities for K-12 students and teachers. The Web site provides Nanoproducts Guides, Worksheets, Teacher Guides, and Activity Pages.

Since a growing number of consumer products employ nanotechnology, the NNIN has assembled a set of activity sheets describing some common and not so common applications of nanotechnology that students can relate to. In general, these products take advantage of some unique property of materials at the nanoscale to provide improved functionality. All the following guides and worksheets are for the following consumer nanotechology products. Some of the consumer product activities include:

- Nanofilm Lens Cleaner- Clarity Fog Eliminator
- Eagle One Nanowax
- Kodak Ultima Picture Paper
- NDMX Golf Balls
- Stain Free "Nano-pants"
- Wilson Double Core Tennis Balls
- Shock Doctor "Hotbed" Insoles
- Nanotech Tennis Racket
- Pilkington Self-Cleaning Glass

Go to: http://www.nnin.org/nnin_k12nanotechproducts.html

READING MATERIALS

Brezina, Corona. *Careers in Nanotechnology*. New York: Rosen Publishing Group, 2007.

Drexler, Eric K., Christine Peterson, and Gayle Pergamit. *Unbounding the Future*. New York: William Morrow and Company, Inc., 1991.

Feynman, Richard P. *The Meaning of it All: Thoughts of a Citizen-Scientist*. New York: Perseus Books, 1998.

Hall, J. Storrs. *Nanofuture: What's Next For Nanotechnology*. Amherst, NY: Prometheus Books, 2005.

Ratner, Mark, and Daniel Ratner. *Nanotechnology: A Gentle Introduction to the Next Big Idea*. Upper Saddle River, NJ: Prentice Hall, 2003.

VIDEOS

Vega Science Trust, England: Videos on basics of nanotechnology and how it will change the world. http://www.vega.org.uk/series/tnbt/nanotechnology/index.php

Work Force Preparation. Are We Prepared to Get Into the Nanotechnology Workforce? Professor Wendy Crone. Conversations in Science. Madison Metropolitan School District. UW-Madison Interdisciplinary Education Group. http://mrsec.wisc.edu/Edetc/cineplex/MMSD/prepared.html

Penn State University. *Amazing Creatures with Nanoscale Features.* This animation is an introduction to microscopy, scale, and applications of nanoscale properties. It introduces some of the tools that are used by scientists to visualize samples that are smaller than what we can see with our eyes. This activity is available for use via the Center Web site at http://www.cneu.psu.edu/edToolsActivities.html

N is for Nanotechnology. NISE Network. "N" is for Nanotechnology" is a 30-minute documentary exploring the hypes, hopes, and facts of this fascinating field as seen through the eyes of award-winning scientists, industry leaders, and writers. http://www.knhproductions.ca/nisnano/trailer.html

Web casts. Presentation. University of Texas, Nanotechnology

"Nanoscience: Big Science at Tiny Scales" Join Dr. Paul Barbara for an exploration of what nanotechnology is and can be, and how researchers at the University of Texas are helping to shape our nanotechnology future. ESI http://streamer.cit.utexas.edu:16080/esi/webcast

WEB SITES

Harvard University: Includes monthly cable casting of NanoTech News via New England Cable News, NanoScale Science & Engineering Presentations, NSEC guest researcher appearances, multimedia research updates, educational collaboration & dissemination activities, Teacher Programs, outreach to students & knowledge transfer to public. http://www.nsec.harvard.edu/education.htm

University of California at Berkeley: Includes researcher presentations, interactive demos and facilitated activities, physical/virtual contextual bridge exhibits, insights into research, human map of research and talking techno heads. http://scienceview.berkeley.edu/wor/Exhibits.html and http://scienceview.berkeley.edu/wor/Partners.htm

University of Chicago: Includes details on research opportunities for undergraduates and high school students, demos for elementary school kids and internships for high school students. Also, includes videos like "Sights & Sounds of Science," "DragonflyTV," etc. Talks about museum collaborations & science camps. http://mrsec.uchicago.edu/outreach/

SOMETHING TO DO

Like a mystery? Solve the Case of the Green Milk by going to The Lawrence Hall of Science, Nanozone site. http://nanozone.org/what.htm

Appendix A

Bibliography

Atkinson, William Illsey. *Nanocosm: Nanotechnology and the Big Changes Coming from the Inconceivably Small.* New York: AMACOM/American Management Association, 2003.

Berube, David M. *Nano-Hype: The Truth behind the Nanotechnology Buzz.* Amherst, NY: Prometheus Books, 2005.

Brezina, Corona. *Careers in Nanotechnology.* New York: Rosen Publishing Group, 2007.

Committee on Implications (author). *Implications of Emerging Micro- and Nanotechnologies.* Washington, DC: National Academies Press, 2002.

Crandall, B.C. *Nanotechnology: Molecular Speculations on Global Abundance.* Cambridge, MA: The MIT Press, 1996.

Crandall, B.C., and James Lewis (Ed.). *Nanotechnology: Research and Perspectives.* Cambridge, MA: The MIT Press, 1992.

Di Ventra, Massimiliano, and Stephane Evoy (Eds.). *Introduction to Nanoscale Science and Technology.* New York: Springer, 2004.

Drexler, Eric K. *Nanosystems: Molecular Machinery, Manufacturing, and Computation.* New York: John Wiley & Sons, 1992.

———. *Engines of Creation.* Garden City, NY: Anchor Press, 1987.

Drexler, Eric K., Christine Peterson, and Gayle Pergamit. *Unbounding the Future.* New York: William Morrow and Company, Inc., 1991.

Feynman, Richard P. *The Character of Physical Law, The 1964 Messenger Lectures.* Cambridge, MA: MIT Press, 1967.

———. *Six Easy Pieces: Essentials of Physics Explained by Its Most Brilliant Teacher.* New York: Perseus Books, 1998.

———. *Six Not So Easy Pieces: Einstein's Relativity, Symmetry and Space-Time.* Reading, MA: Addison Wesley, 1997.

———. *The Meaning of it All: Thoughts of a Citizen-Scientist.* New York: Perseus Books, 1998.

Fishbine, Glenn. *The Investor's Guide to Nanotechnology & Micromachines.* New York: John Wiley & Sons, 2001.

Foster, Lynn E. *Nanotechnology: Science, Innovation, and Opportunity.* Upper Saddle River, NJ:Prentice Hall, 2005.

Fouke, Janie (Ed.). *Engineering Tomorrow: Today's Technology Experts Envision the Next Century*. Piscataway, NJ: IEEE Press, 2000.

Freitas, Jr., Robert A, and Ralph C. Merkle. *Kinematic Self-Replicating Machines.* Austin, TX: Landes Bioscience, 2004.

Freitas, Jr., Robert A. *Nanomedicine: Basic Capabilities, Vol. 1.* Austin, TX: Landes Bioscience, 1999.

Fritz, Sandy. *Understanding Nanotechnology: From the Editors of Scientific American.* New York: Warner Books, 2002.

Fujimasa, Iwao. *Micromachines: A New Era in Mechanical Engineering.* Oxford, NY: Oxford University Press, 1996.

Goodsell, D.S. *BioNanotechnology: Lessons from Nature.* Hoboken, NJ: Wiley-Liss, Inc., 2004.

Gross, Michael. *Travels to the Nanoworld: Miniature Machinery in Nature and Technology.* New York: Perseus Books Group, 2001.

Hall, J. Storrs. *Nanofuture: What's Next for Nanotechnology.* Amherst, NY: Prometheus Books, 2005.

Hamakawa, Yoshihiro. *Thin-Film Solar Cells.* New York: Springer-Verlag, 2004.

Imry, Yoseph. *Introduction to Mesoscopic Physics.* Oxford, NY: Oxford University Press, 2002.

Johnson, Rebecca, L. *Nanotechnology (Cool Science).* Minneapolis, MN: Lerner Publications, 2005.

Jones, R.L. *Soft Machines: Nanotechnology and Life.* Oxford, UK: Oxford University Press, 2004.

Jones, M. Gail, Michael R. Falvo, Amy R. Taylor, and Bethany P. Broadwell. *Nanoscale Science.* Arlington, VA: NSTA Press, 2007.

Karn, Barbara, Vicki Colvin, Paul Alivasatos, and Tina Masciangioli (Eds.). *Nanotechnology and the Environment.* Washington, DC: American Chemical Society, 2004.

Krummenacker, Markus, and James J. Lewis. *Prospects in Nanotechnology: Toward Molecular Manufacturing.* New York: John Wiley & Sons, 1995.

Luryi, Serge, and Jimmy Xu. *Future Trends in Microelectronics: Reflections on the Road to Nanotechnology.* Boston: Kluwer Academic Publishers, 1996.

Mulhall, Douglas. *Our Molecular Future: How Nanotechnology, Robotics, Genetics, and Artificial Intelligence will Transform Our World.* Amherst, NY: Prometheus Books, 2002.

National Academy Press (author). *Small Wonders, Endless Frontiers: A Review of the National Nanotechnology Initiative.* Washington, DC: National Academy Press, 2002.

Newton, David E. *Recent Advances and Issues in Molecular Nanotechnology.* Westport, CT: Greenwood Press, 2002.

Poole, Charles P., and Frank J. Owens. *Introduction to Nanotechnology.* Hoboken, NJ: John Wiley & Sons, Wiley-Interscience, 2003.

Ratner, Mark, and Daniel Ratner. *Nanotechnology: A Gentle Introduction to the Next Big Idea.* Upper Saddle River, NJ: Prentice Hall, 2003.

Ratner, Daniel, and Mark A. Ratner. *Nanotechnology and Homeland Security New Weapons for New Wars.* Upper Saddle River, NJ: Prentice Hall, 2003.

Regis, Edward. *Nano: The Emerging Science of Nanotechnology.* Boston, MA: Little Brown & Co., 1996.

———. *Nano! The True Story of Nanotechnology—the Astonishing New Science That Will Transform the World.* London, United Kingdom: Transworld Publishers Ltd., 1997.

Rietman, Edward A. *Molecular Engineering of Nanosystems Series: Biological and Medical Physics, Biomedical Engineering.* New York: Springer, 2001.

Roco, Mihail C., and William Sims Bainbridge. *Societal Implications of Nanoscience and Nanotechnology.* Boston, MA: Kluwer Academic Publishers, 2001.

———. (Eds.). *Converging Technologies for Improving Human Performance: Nanotechnology, Biotechnology, Information Technology and Cognitive Science.* New York: Springer, 2004.

Sargent, Ted. *The Dance of the Molecules: How Nanotechnology is Changing Our Lives.* New York: Thunder's Mouth Press, 2006.

Schulte, Jurgen (Ed.). *Nanotechnology: Global Strategies, Industry Trends and Applications.* New York: John Wiley & Sons, 2005.

Scientific American (authors). *Key Technologies for the 21st Century: Scientific American: A Special Issue.* New York: W.H. Freeman & Co, 1996.

———. (author). *Understanding Nanotechnology.* New York: Grand Central Publishing, 2002.

Shelley, Toby. *Nanotechnology: New Promises, New Dangers (Global Issues).* London, UK: Zed Books, 2006.

Smalley, R.E. *Carbon Nanotubes: Synthesis, Structure, Properties and Applications.* New York: Springer, 2001.

Uldrich, Jack. *Investing in Nanotechnology: Thank Small. Win Big.* Cincinnati, OH: F + W Publications, 2006.

Uldrich, Jack, and Deb Newberry. *Next Big Thing Is Really Small: How Nanotechnology Will Change the Future of Your Business.* New York: Crown Publishing Group, 2003.

Wiesner, Mark, and Jean-Yves Bottero. *Environmental Nanotechnology.* New York. McGraw Hill Professional Publishing, 2007.

Appendix B

Companies in Nanotechnology Research and Development

AccuFlex® Evolution Golf Shaft by Accuflex®: http://www.accuflexgolf Designers and manufacturers of a new nanoparticle golf shaft. http://www.accuflexgolf.com

Advanced Magnetics: http://www.advancedmagnetics.com/ Advanced Magnetics, Inc. is a developer of iron oxide nanoparticles used in pharmaceutical products.

Agilent Technologies: http://www.home.agilent.com Agilent Technologies offers a wide range of high-precision atomic force microscopes (AFM).

Alnis Biosciences, Inc.: http://www.alnis.com Alnis is a drug development company with a therapeutic technology to treat cancer as well as infectious and inflammatory diseases.

Altairnano: http://www.altairnano.com/ Altairnano is an industry in the development and manufacturing of ceramic nanomaterials.

Angstrom Medica, Inc.: http://www.angstrommedica.com/ Angstrom Medica is a life-science biomaterials company that implements nanotechnology for orthopedic applications.

Anpath Group, Inc. formerly EnviroSystems: http://www.envirosi.com/ A company that develops products designed to prevent infectious diseases without harming the environment.

Applied Materials, Inc.: http://www.appliedmaterials.com/about/index.html Applied Materials, Inc. provides, service and software products for the fabrication of semiconductor chips, flat panels, and solar photovoltaic cells.

Applied Nanoworks: http://www.appliednanoworks.com/ Applied NanoWorks is a technology company that produces additives for the coatings, plastics, ink, and adhesives industries.

Arryx Inc.: http://www.arryx.com/about.html Provides technology that uses focused light that function like microscopic tweezers to grab small objects.

Asia Pacific Fuel Cell Technologies: http://www.apfct.com/ The company is involved with PEM fuel cell technology.

Aspen Aerogels: http://www.aerogel.com/ A manufacturer of aerogels in easy-to-apply blankets and fabricated packages, two to eight times more effective than existing insulations.

BASF, The Chemical Company: http://www.corporate.basf.com Currently produces several nanoproducts, such as nanoparticle pigments and nanoscale titanium dioxide particles.

Bayer: http://www.bayer.com/en/Profile-and-Organization.aspx Bayer is a research-based, growth-oriented global enterprise with core competencies in the fields of health care, nutrition, and high-tech materials. Bayer is successfully exploiting the innovative potential of nanotechnology.

BioDelivery Sciences International: http://www.biodeliverysciences.com A pharmaceutical company that has patented drug delivery technologies.

Bell Labs, Alcatel-Lucent: http://www.alcatellucent.com/wps/portal/BellLabs New Jersey Nanotechnology Consortium (NJNC) is created by the State of New Jersey, the New Jersey Institute of Technology, and Lucent Technologies to conduct research, develop nanotechnology prototype devices, and to train more nanotech scientists and specialists.

Babolat® NS™ Drive Tennis Racket by Babolat®: http://www.babolat.com/english/tennis/technology/index.php?src=tennis The company produces carbon nanotubes used to stiffen key areas of the racquet head and shaft.

Cabot Corporation: http://w1.cabot-corp.com/controller.jsp?N=21+1000 This company produces nano-structured and submicron particles that are used in the production of tires, inks, and plastics.

Carbon Nanotechnologies, Inc.: http://www.cnanotech.com CNI is a producer of Buckytubes.

Carter Wallace. First Response® Home Pregnancy Test by Carter-Wallace®: http://nanopedia.case.edu/NWPage.php?page=pregnancy.tests The home pregnancy test is an example of how nanoproperties can be used.

Cima Nanotech, Inc.: http://www.cimananotech.com/products.aspx Cima NanoTech manufactures nanomaterials-based products for use in electronics applications.

CG2 NanoCoatings, Inc.: http://cg2nanocoatings.com NanoCoatings provides innovative and practical nanotechnology-based coating solutions.

DayStar: http://www.daystartech.com/ DayStar's goal is to provide low-cost, high-volume solar-electric photovoltaic (PV) cells.

Dendritech: http://www.dendritech.com/ The company provides antibody conjugates used in an immunoassay for rapid and sensitive detection of markers indicative of heart attacks.

Dendritic Nanotechnologies, Inc.: http://dnanotech.com/ DNT, Inc. and the National Cancer Institute (NCI) have entered into a Small Business Innovation Research (SBIR) contract to develop a new generation of targeted diagnostic and therapeutic delivery technology for the early detection and treatment of epithelial ovarian cancer.

Dockers® Go Khaki® by Dockers®: http://www.popsci.com/popsci/science/92a80b4511b84010vgnvcm1000004eecbccdrcrd.html The clothing company offers pants which promise to keep your legs stain-free using nanotechnology."

DuPont Titanium Technologies: http://www.titanium.dupont.com/ DuPont has been a pioneer in titanium dioxide technology for the coatings industry.

Eagle One Nanowax® by Eagle One: http://www.eagleone.com/pages/products/product.asp?itemid=1103&cat=5010

NanoWax uses nanotechnology to fill in these scratches and conceal the swirl marks.

Eastman Kodak, Ultima® Photo Paper by Eastman Kodak® Company: http://nanotechwire.com KODAK Ultima Picture Paper employs ceramic nanoparticles to resist the effects of heat, humidity, light, and ozone.

Ecologycoatings Liquid Nanotechnology: http://www.ecologycoatings.com The company uses nano-sized particles of mineral oxides to create waterproof coatings for paper.

EnviroSystems: http://www.envirosi.com The company's goal is to produce products designed to prevent infectious diseases without harming the environment.

Evident Technologies: http://www.evidenttech.com The company develops products based on proprietary quantum dot technology.

eSpin Technologies: http://www.espintechnologies.com/company.htm The company manufactures Nanofibers used in industries such as aerospace, health care, and energy storage.

FEI Company: http://fei.com FEI designs and manufactures a wide variety of electron microscopes and nanotechnology tools and components.

Filtration Technology, Inc.: http://www.filtrationtechnology.com/foodchem.shtml

Filtration Technology, Inc., are filtration and contamination control specialists that provide cleanroom facilities and air and liquid filtration.

General Motors® Automotive Exterior by General Motors® Inc.: http://www.azonano.com

GM is now using about 660,000 pounds of nanocomposite material per year.

Goodweaver: http://www.goodweaver.com/ Goodweaver produces nanosilver antibacterial and deodorant insole that stops itching and prevents athletic foot.

Helix Material Solutions: http://www.helixmaterial.com/ HMS provides single-walled carbon nanotubes (SWNT) and a variety of multiwalled carbon nanotubes (MWNT) for applications in electronics, biology, and medicine.

Hewlett Packard: http://www.hpl.hp.com/research/about/nanotechnology.html Researchers at Hewlett-Packard are focusing on the fabrication of nanometer-scale structures.

Hitachi High-Technologies Corporation: http://www.hitachi-hitec.com/global/ Hitachi High-Technologies is an integrated organization that develops, manufactures, markets, and services equipment and systems in the emerging field of nanotechnology.

Hyperion Catalysis International: http://www.fibrils.com Hyperion Catalysis International is a company in carbon nanotube development and commercialization.

IBM Research: http://researchweb.watson.ibm.com/nanoscience/ IBM uses scanning tunneling microscopy (STM) and atomic force microscopy (AFM) as structural probes, and, along with electron beam lithography, as tools for the modification of materials at the atomic and nanometer scales.

Innovative Skincare: http://innovativeskincare.com/ Innovative® Skincare produces SPF 20 Sunscreen Powder that provides the ultimate broad-spectrum sun protection.

Intel: http://www.intel.com/ Develop the first 45nm processor. Intel entered the nanotechnology era in 2000 when it began volume production of chips with sub-100nm length transistors.

Isotron: http://www.isotron.net/isotron.php?style=gold The company employs protective coatings for equipment and decontamination technologies for homeland security, industrial, and military applications.

Kereos Targeted Therapeutics and Molecular Imaging: http://www.kereos.com/ Kereos Inc. develops targeted therapeutics and molecular imaging agents designed to detect and treat cancer and cardiovascular disease early.

Konarka Technologies: http://www.konarkatech.com/ Konarka builds a light-activated power plastic that converts light to energy—anywhere. The light-activated power plastic is inexpensive, lightweight, flexible, and versatile.

Lands' End: ThermaCheck® Vest by Lands' End® Inc. http://www.landsend.com/ The company applies nanotechnology to each fiber in the fleece that resists static cling and shock, repels lint, and pet hair.

Luna nanoWorks: http://www.lunananoworks.com/ Luna nanoWorks produces high quality proprietary carbon nanomaterials, including single-walled carbon nanotubes.

Mercedes-Benz: http://www.mercedes-benz.com The automobile company introduced nanoparticle new paint system for its Mercedes models.

MetaMateria Partners:www.metamateria.com MetaMateria Partners is a developer and manufacturer of nanopowders and ceramic manufacturing methods for energy storage such as batteries and capacitor.

Metal Nanopowders: http://www.metalnanopowders.com/ The company produces metal nanopowders that can be used in magnetic tapes and targeted drug delivery.

Molecular Imprints: http://www.molecularimprints.com Molecular Imprints design, develop, manufacture, and support imprint lithography systems to be used by semiconductor device and other industry manufacturers.

Motorola® Organic Light Emitting Diodes (OLEDs) by Motorola®: www.motorola.com The research arm of Motorola Inc. has built a 5-inch color video display prototype using its own carbon nanotube (CNT) technology.

mPhase and Lucent Technologies: http://www.mphasetech.com mPhase and Lucent Technologies are laboratories that are developing new kinds of batteries.

Nanergy Inc.: http://www.nanergyinc.com/ Nanergy is a company that is focus on products that harness nanotechnology, and especially photovoltaic nanofilms.

Nano-C Inc.: http://www.nano-c.com Nano-C is a developer in the production of high-quality fullerenic materials including C60, C70, C84, and fullerene black.

Nanochem: http://www.nanochem.com NanoChem has developed a new chemical process that efficiently removes ammonia from wastewaters.

Nanoco Technologies: http://www.nanocotechnologies.com Nanoco Technologies manufactures fluorescent nanocrystals from semiconductor and metallic materials known as quantum dots.

Nanofilm Technology: http://www.nanofilmtechnology.com Nanofilm develops optically clear, thin coatings, self-assembling nanolayers, and nanocomposites that act as a protective layer for displays, such as computer displays and cell phone windows.

NanoDynamics: http://www.nanodynamics.com NanoDynamics provides nano-enabled solutions in the fields of energy, automotive, water processing, and consumer products.

NanoHorizons: http://nanohorizons.com/ A company that provides anti-odor/antimicrobial protection to natural and synthetic fibers and fabrics.

NanoInk: http://www.nanoink.net/ The company's product is The NSCRIPTOR™, which allows scientists to perform experiments using Dip Pen Nanolithography®.

Nano Interface Technology, Inc.: http://www.nanointerfacetech.com Nano Interface Technology is a research organization committed to developing nanotechnologies in biotechnology and the drug delivery areas.

Nanolab: http://www.nano-lab.com Nanolab is a manufacturer of carbon nanotubes and a developer of nanoscale devices.

Nanoledge: http://nanoledge.com Nanoledge provides carbon nanotubes-filled resins for high-performance equipment applications that include water sports, motor sports, and balls sports, including tennis rackets.

NanoMedica, Inc.: http://www.nanomedica.com/ The company is developing therapeutics for the treatment of cancer using nanotechnology.

Nanonex Corp: http://www.nanonex.com Manufacturer of imprint, step-and-flash lithography systems.

NanoOpto Corporation: http://www.nanoopto.com NanoOpto's products have applications in optical communications, digital imaging, and optical storage (CD/DVD drives).

Nanostellar: http://www.nanostellar.com/home.htm Nanostellar is developing technology for applications that include automobile catalysts, petrochemical industry, displays, fuel cells and batteries.

Nanosolar: http://www.nanosolar.com Nanosolar has developed proprietary technology that makes it possible to simply roll-print solar cells.

NanoSonic, Inc.: http://www.nanosonic.com NanoSonic, Inc. develops revolutionary new molecular self-assembly processes that allow the controlled synthesis of material structure at the nanometer level and the manufacturing of new materials. Products include Metal Rubber and sensors and systems.

Nanophase Technologies: http://www.nanophase.com/ Nanophase is developing processes for the commercial manufacturing of nanopowder metal to be used in fuel cells, personal care, and wood preservation.

Nanospectra Biosciences, Inc.: http://www.nanospectra.com/ Nanospectra Biosciences, Inc. is developing a therapeutic medical device to destroy solid tumors.

Nanosys: http://www.nanosysinc.com The company is using thin-film electronics technology to help develop high performance fuel cells for use in portable consumer electronics such as laptop computers and cell phones.

Nanosurf: http://www.nanosurf.com Nanosurf develops scanning probe microscopes (SPM)

Nanotec. Nanoprotect Glass and Automotive Glass by Nanotec Pty Ltd.: http://www.nanotec.com.au/ Nanoprotect Glass is a special nanotechnology product with a long-term self-cleaning effect for glass and ceramic surfaces.

Nano-Tex® Textiles by Nano-Tex®, Inc., LLC.: http://www.nano-tex.com Nano-Tex™'s technology fundamentally transformed fabric at the nano-level to improve everyday clothing by being stain-resistant and static free.

NEC: http://www.nec.com NEC is developing a carbon nanohorn electrode that it claims can create 20 percent more power in fuel cells used in laptop computers and mobile phones.

Nucryst: http://www.nucryst.com Nucryst Pharmaceuticals develop medical products that fight infection and inflammation based on its patented nanocrystalline silver technology.

OptoTrack, Inc.: http://www.optotrack.com/ot_info/index.html Optotrack, Inc. is a company in the development of optical, imaging, biological sensors, and microsystems based on microfabrication and nanotechnology.

Owlstone Nanotech: http://www.owlstonenanotech.com Owlstone has created a complete chemical detection system a hundred times smaller and a thousand times cheaper than other currently available devices.

Oxonica: http://www.oxonica.com/index.php Oxonica Energy division is focused on the development of ENVIROX™, a diesel fuel-borne catalyst which reduces fuel consumption and exhaust emissions.

Pacific Nanotechnology: http://www.pacificnanotech.com Pacific Nanotechnology provides atomic force microscope products and services.

Pilkington: http://www.pilkingtonselfcleaningglass.co.uk Pilkington Activ™ is the world's first self-cleaning glass to use a microscopic coating with a unique dual action.

Powdermet, Inc.: http://www.powdermetinc.com Powdermet manufactures metal and ceramic nano-engineered fine powders and particulates used in many fields including electronic materials, metal cutting, and paints.

Proctor & Gamble: http://www.pg.com/en_US/index.jhtml Proctor and Gamble have been using nano-structured fluids research in consumer products.

pSivida: http://www.psivida.com/ pSivida is a bio-nanotechnology company that has developed drug delivery products in the health-care sector.

QuantumSphere, Inc.: http://www.qsinano.com QuantumSphere Develops catalyst materials and electrode devices for clean-energy applications such as batteries and micro-fuel cells for portable power.

Rosseter Holdings Ltd.: http://www.e-nanoscience.com. Rosseter Holdings Limited, located in Cyprus, is a company that specializes in large-scale production of carbon nanotubes and related materials.

Samsung® Refrigerator by Samsung®: http://www.samsung.com Samsung utilized nanotechnology and applied it to the interior coating of their refrigerators for effective sterilization, deodorization, and antibacterial effects.

Shemen Industries.Canola Active Oil by Shemen Industries: http://goliath.ecnext.com/ The Canola Active Oil is used to inhibit the transportation of cholesterol from the digestive system into the bloodstream.

Smith and Nephew: http://www.accuflexgolf.com/tech.asp) Acticoat® Wound Dressings by Smith & Nephew provides antimicrobial barrier protection.

Sharper Image, Antibacterial Silver Athletic and Lounging Socks by Sharper Image: http://www.sharperimagebest.com/zn021.html) The quarter-length sports socks are knitted with a cotton material containing millions of invisible silver nanoparticles.

Solaris Nanosciences Corporation: http://www.solarisnano.com/contactus.php The Solaris primary goal is to provide low manufacturing cost, high-efficiency, and long-life solar cells in the projected growth of the global renewable energy market.

Sony® Corporation Organic Light Emitting Diodes (OLEDs) by Sony® Corporation: http://www.sony.com/ Made of nano-structured polymer films, the OLED screens emit their own light and are lighter, smaller, and more energy efficient than conventional liquid crystal displays.

T-2® Photocatalyst Environment Cleaner by T-2®: http://www.t-2.biz/index.php?cf=t-2_intro The company has developed a surface cleaner that can cause the breakdown of organic toxins, odors, and more.

Toshiba, Lithium-Ion Battery by Toshiba Corporation: http://www.toshiba.co.jp/about/press/2005_03/pr2901.htm The company is developing a new kind of battery that contains nanoparticles that can quickly absorb and store vast amount of lithium ions, without causing any deterioration in the electrode.

Veeco: http://www.veeco.com Veeco is a leader in atomic force and scanning probe microscopy. Vecco produces scanning probe microscope systems (AFM/SPM).

Wilson Sporting Goods: http://www.wilson.com Wilson announced it was the first golf equipment manufacturer to strategically use nanotechnology to develop stronger and lighter materials in their sporting equipment.

Xintek, Inc.: http://www.xintek.com Xintek's major products include carbon nanotube (CNT)-based field emission electron source, field emission grade CNT material, field emission x-ray source, and CNT AFM probes.

Zyvex: http://www.zyvex.com/ Zyvex develops nanomanipulation products that are compatible with scanning electron microscopes and standard optical microscopes.

Appendix C

Nanotechnology Web sites

Please note that the author has made a consistent effort to include up-to-date Web sites. However, over time, some Web sites may no longer be posted.

GOVERNMENT AGENCIES

Department of Agriculture (DOA): The Department of Agriculture's multifaceted mission is to ensure a safe food supply; care for agricultural lands, forests, and rangelands,. http://www.usda.gov/wps/portal/usdahome

Department of Defense (DOD): The mission of the Department of Defense is to provide the military forces needed to deter war and to protect the security of the United States. http://www.defenselink.mil/

Department of Education (DOE): Its original directive remains its mission today—to ensure equal access to education and to promote educational excellence throughout the nation. http://www.ed.gov/index.jhtml

Department of Energy: The Department of Energy's (DOE) goal is to advance the national, economic, and energy security of the United States. http://www.energy.gov/index.htm

Department of Homeland Security(DHS): The Department of Homeland Security is a federal agency whose primary mission is to help prevent, protect against, and respond to acts of terrorism on United States soil. http://www.dhs.gov/index.shtm

Environmental Protection Agency(EPA): The Environmental Protection Agency was created in 1970 and was established in response to growing public concern about unhealthy air, polluted rivers and groundwater, unsafe drinking water, endangered species, and hazardous waste disposal. http://www.epa.gov/

Food and Drug Administration (FDA). The U.S. Food and Drug Administration regulates a wide range of products, including foods, cosmetics, drugs, devices, and veterinary products, some of which may utilize nanotechnology or contain nanomaterials. http://www.fda.gov/nanotechnology/

Los Alamos National Laboratory: The Laboratory has served the nation by developing and applying the best science and technology to ensure its safety and security. http://www.lanl.gov/

National Aeronautics and Space Administration (NASA): NASA's mission is to pioneer the future in space exploration, scientific discovery, and aeronautics research. http://www.nasa.gov/home/index.html

National Institute for Occupational Safety and Health (NIOSH): The agency is responsible for developing and enforcing workplace safety and health regulations. http://www.cdc.gov/niosh/homepage.html

National Institute of Standards and Technology (NIST): NIST's mission is to promote U.S. innovation and industrial competitiveness by advancing measurement science, standards, and technology in ways that enhance economic security and improve our quality of life. http://www.nist.gov/public_affairs/nanotech.htm

National Nanotechnology Initiative (NNI): The National Nanotechnology Initiative (NNI) is a federal R&D program established to coordinate the multiagency efforts in nanoscale science, engineering, and technology. http://www.nano.gov/

National Renewable Energy Laboratory (NREL): The nation's primary laboratory for renewable energy and energy efficiency research and development (R&D). www.nrel.gov/

National Science Foundation (NSF): Their goal is to promote the progress of science; to advance the national health, prosperity, and welfare; to secure the national defense. http://www.nsf.gov/crssprgm/nano/

National Science and Technology Council (NSTC): A primary objective of the NSTC is the establishment of clear national goals for Federal science and technology investments http://www.ostp.gov/nstc/index.html

Sandia National Laboratories: Sandia National Laboratories has developed science-based technologies that support our national security. http://www.sandia.gov/

OTHER WEB SITES

All Things Nano: Museum of Science, Boston. http://antill.com/MOS/

AZoNano Information: The aim of AZoNano.com is to become the primary Nanotechnology information source for the science, engineering, and design community worldwide. http://www.azonano.com/aboutus.asp

Big Picture on NanoScience: The Wellcome Trust is an independent charity funding research to improve human and animal health. http://www.wellcome.ac.uk/node5954.html

CMP Cientifica: News, networks, conferences, and nanotechnology resources in Europe. http://www.cientifica.com

Foresight Institute: Nonprofit institute focused on nanotechnology, the coming ability to build materials and products with atomic precision, and systems to aid knowledge exchange and critical discussion, thus improving public and private policy decisions. http://www.foresight.org/

Foresight Institute, Nanomedicine: Nanomedicine may be defined as the monitoring, repair, construction and control of human biological systems at the molecular level, using engineered nanodevices and nanostructures. Includes the Nanomedicine Art Gallery. http://www.foresight.org/Nanomedicine

How Stuff Works: How Nanotechnology Will Work: Animated narrative shows how Nanotechnology has the potential to totally change manufacturing, health care and many other areas. http://www.howstuffworks.com/nanotechnology.htm

IBM Almaden STM Molecular Art: Some of the famous images of atoms and molecules made with IBM's scanning tunneling microscope. http://www.almaden.ibm.com/vis/stm/lobby.html

Institute of Nanotechnology (UK): The Institute of Nanotechnology has been created to foster, develop, and promote all aspects of science and technology in those domains where dimensions and tolerances in the range of 0.1 nm to 100nm play a critical role. http://www.nano.org.uk/

NanoBusiness Alliance: Provides nanobusiness information. www.nanobusiness.org/

Nanogloss: Online dictionary of nanotechnology. http://www.nanogloss.com

Nanooze: A science magazine about nanotechnology for kids. www.nanooze.org

NanoScale Science Education: NanoScale Science Education Research Group that offers K-12 nanotechnology materials. http://ced.ncsu.edu/nanoscale/nanoteched.htm

Nano Science and Technology Institute: The Nano Science and Technology Institute (NSTI) is chartered with the promotion and integration of nano and other advanced technologies through education, technology and business development. http://www.nsti.org/

NanoSpace: The Center for NanoSpace Technologies is a Texas-based nonprofit scientific research and education foundation. http://www.nanospace.org

Nanotechnology: Institute of Physics monthly journal for aspects of nanoscale science and technology. http://www.iop.org/journals/nano

Nanotechnology Now: Provides introduction to nanotechnology, general information, images, interviews, news, events, research, books, glossary, and links. http://nanotech-now.com/

National Cancer Institute: Provides information about common cancer types. http://www.cancer.gov/

National Heart, Lung, and Blood Institute: The National Heart, Lung, and Blood Institute (NHLBI) provides leadership for a national program in diseases of the heart, blood vessels, lung, and blood; blood resources; and sleep disorders. http://www.nhlbi.nih.gov/index.htm

NIST. Scanning Tunneling Microscope (STM): Describes the invention of the topografiner, a precursor instrument, between 1965 and 1971, and also tells of the STM's development. http://physics.nist.gov/GenInt/STM/stm.html

Science Friday Kids' Connection in association with Kidsnet: http://www.sciencefriday.com/kids/sfkc20021206-1.html

Scientific American: Nanotechnology articles, some free and some archived. http://www.sciam.com/nanotech/

Secret Worlds: The Universe Within. View the Milky Way at 10 million light years from the Earth. Then move through space toward the Earth in successive orders of magnitude until you reach a tall oak tree. http://micro.magnet.fsu.edu/primer/java/scienceopticsu/powersof10/

Small Times: Daily articles covering MEMS, nanotechnology, and microsystems, with a business angle. http://www.smalltimes.com

The Incredible Shrunken Kids: http://www.sciencenewsforkids.org/articles/20040609/Feature1.asp

The NanoTechnology Group: A consortium of nano companies, universities, and organizations developing a nano science curriculum for K-12. http://www.TheNanoTechnologyGroup.org

The Nanotube Site: Michigan State University's Library of Links for the Nanotube Research Community. http://www.pa.msu.edu/cmp/csc/NTSite/nanopage.html

Vega Science Trust, England: Videos on basics of nanotechnology and how it will change the world. http://www.vega.org.uk/series/tnbt/nanotechnology/index.php

Virtual Journal of Nanoscale Science & Technology: A weekly multijournal compilation of the latest research on nanoscale systems. http://www.vjnano.org

Xerox/Zyvex Nanotechnology: Former Xerox Palo Alto Research Center Nanotechnology page: brief introduction to core concepts of molecular nanotechnology (MNT), and links for further reading. http://www.zyvex.com/nano/

Zyvex.com: Provides a brief introduction to the core concepts of molecular nanotechnology. http://www.zyvex.com/nano/

Appendix D

Nanotechnology Videos and Audios

The following videos and audios are suggested to enhance your understanding of nanotechnology topics and issues. Some of the videos listed are presented in the chapters. However, you may wish to review other videos as well.

Please note that the author has made a consistent effort to include up-to-date Web sites. However, over time, some Web sites may move or no longer be posted.

Viewing some of these videos may require special software called plug-ins. Therefore, you will need to download certain software to view the videos. You may need to upgrade your player to the most current version.

VIDEO

How Breast Cancer Spreads. Sutter Health. This group has a number of cancer videos. http://cancer.sutterhealth.org/information/bc_videos.html

Faster Results for Breast Cancer. Pathologists Use Digital Imaging to Speed up Cancer Diagnosis. Science Daily. http://www.sciencedaily.com/videos/2006-02-06/

Detecting Deadly Chemicals. Science Daily. A sampler gun that can now be used to detect harmful or dangerous diseases such as anthrax. http://www.sciencedaily.com/videos/2006-12-10/

Work Force Preparation. Are We Prepared to Get into the Nanotechnology Workforce? Professor Wendy Crone. Conversations in Science. Madison Metropolitan School District. UW-Madison Interdisciplinary Education Group. http://mrsec.wisc.edu/Edetc/cineplex/MMSD/prepared.html

Video Game: PlayGen develops games for learning and teaching. NanoMission™ is developing first scientifically accurate interactive 3d learning games based on understanding nanosciences and nanotechnology. To see a sample of one of their games with a theme of nanoscience and cancer go to: http://www.playgen.com/home/content/view/30/26/

Scanning Probe Microscopy. Professor Wendy Crone The next big thing or smaller. Conversations in Science. Madison Metropolitan School District. UW-Madison Interdisciplinary Education Group. http://mrsec.wisc.edu/Edetc/cineplex/MMSD/scanning2.html

Stroke Stopper. Neuroradiologists Treat Brain Strokes with New Kind of Stent, Science Daily Video. Go to: http://www.sciencedaily.com/videos/2006-04-07/

Nanoscale. Professor Wendy Crone What is a nanoscale. Discusses Quantum Effects, and Quantum Dots. Surface to Volume Ration Makes a Difference is also discussed. Conversations in Science. Madison Metropolitan School District. UW-Madison Interdisciplinary Education Group. http://mrsec.wisc.edu/Edetc/cineplex/MMSD/index.html

When Things Get Small. Google Video. Describes how small is a nanometer? The film showing how scientists layer atoms to form nanodots. http://video.google.com/videoplay?docid=-215729295613330853

Try the Simulator. To see a simulation of a scanning tunnel microscope go to: http://nobelprize.org/educational_games/physics/microscopes/scanning/

Using Nanoscience to understand the properties of matter. Explore Materials. http://www.wpsu.org/nano/lessonplan_detail.php?lp_id=21

What is Matter? http://www.wpsu.org/nano/lessonplan_detail.php?lp_id=21

What is a Molecule? http://www.wpsu.org/nano/media/Molecule.mov

Taking Pictures of What You Can't See. http://www.wpsu.org/nano/media/TakingPictures.mov

The Threat of Bird Flu. ScienceDaily http://www.sciencedaily.com/releases/2007/04/070416092206.htm

NASA Space Elevator. Can we build a 22,000-mile-high cable to transport cargo and people into space? http://www.pbs.org/wgbh/nova/sciencenow/3401/02.html

Forming Carbon Nanotubes. University of Cambridge. Two videos show how nickel reacts in a process called catalytic chemical vapor deposition. This film demonstrates one of several methods of producing nanotubes. Text accompanies the video for better understanding of the process. http://www.admin.cam.ac.uk/news/special/20070301/?

Videos from the Hitachi Corporation. What's Next in Nanotechnology? http://www.hitachi.com/about/corporate/movie/

The Lemelson Center. Nobel laureate William Phillips levitates a magnet to explain how atoms form bonds. There are several videos in this collection. Select the one on levitation. You may be also interested in the other videos as well. http://invention.smithsonian.org/centerpieces/ilives/phillips/phillips.html

CSI: X-Ray Fingerprints. Stimulating atoms to reveal chemicals on fingerprints, http://www.sciencedaily.com/videos/2006-12-08/

What is Nanotechnology? University of Wisconsin-Madison engineer Wendy Crone is on a mission. She and her interns are creating user-friendly exhibits to teach the public about the nanoworld. http://www.sciencedaily.com/videos/2006-06-11/

Exploring the Nanoworld: Movies of nano-structured materials: ferrofluids, memory metals, amorphous metals, LEDs, self-assembly, DNA, Magnetic Resonance Imaging,

Lego models. National Science Foundation supported Materials Research Science and Engineering Center on Nanostructured Materials and Interfaces at the University of Wisconsin-Madison. http://mrsec.wisc.edu/edetc

A Nanotechnology Super Soldier Suit. Nanotechnology in the military. http://www.youtube.com/watch?v=-pNbF29l9Zg&mode=related&search=

Nanokids-Bond With Me. You Tube. http://www.youtube.com/watch?v=R6_vWWGjt10

Monitoring Blood Glucose Without Pain or Blood. A short film of Professor Paranjape and the lab where the diabetes biosensor device is produced. http://college.georgetown.edu/research/molecules/14887.html

Secret Worlds: The Universe Within. Visit the subatomic universe of electrons and protons and viewing the Milky Way. http://micro.magnet.fsu.edu/primer/java/scienceopticsu/powersof10/

G Living. The Phoenix Electric Nano Battery SUV. http://www.youtube.com/watch?v=w-Zv5RFgmWY&NR

National Geographic. National Geographic has five video shots on Nanotechnology. They include the following titles: *Smaller than Small, Land of the Giants, Nano in Nature, Growing Technology, and Stand by for a Change.* This activity is available at: http://www7.nationalgeographic.com/ngm/0606/feature4/multimedia.html

Penn State University. *Amazing Creatures with Nanoscale Features*: This animation is an introduction to microscopy, scale, and applications of nanoscale properties. It introduces some of the tools that are used by scientists to visualize samples that are smaller than what we can see with our eyes. This activity is available for use via the Center Web site at http://www.cneu.psu.edu/edToolsActivities.html

Electron-Beam Lithography. Nanopolis Online Multimedia Library. Electron-beam lithography is a technique for creating extremely fine patterns required for modern electronic circuits. http://online.nanopolis.net/viewer.php?subject_id=139

Is Nanotechnology Going to be The Next Industrial Revolution? Conversations in Science. Madison Metropolitan School District. UW-Madison Interdisciplinary Education Group. http://mrsec.wisc.edu/Edetc/cineplex/MMSD/nano5.html

Cosmetics. Nanopolis Online Multimedia Library. The cosmetics industry was one of the first industries to employ nanotechnology for cosmetics that include creams, moisturizers, and sunscreens. http://online.nanopolis.net/viewer.php?subject_id=274

Carbon Nanotube Transistors. Nanopolis Online Multimedia Library. The carbon nanotubes are ideal building blocks for molecular electronics. http://online.nanopolis.net/viewer.php?subject_id=268

University of Virginia Materials Research Science and Engineering Center. Paladin Pictures, Inc. and the University of Virginia Materials Research Science & Engineering Center (MRSEC). The film, Nano Revolution, was filmed largely on-location at the University of Virginia, and is aimed at introducing people to the field of Nanotechnology. http://www.paladinpictures.com/nano.html.

Dendritic Polymer Adhesives for Corneal Wound Repair Presented by: Mark W. Grinstaff, Ph.D., Metcalf Center for Science and Engineering. http://www.blueskybroadcast.com/Client/ARVO/

Nanoparticles—Chemistry, Structure and Function

Presented by: Karen L. Wooley, Ph.D. and Professor, Washington University in Saint Louis, Department of Chemistry. http://www.blueskybroadcast.com/Client/ARVO/

Introduction to Nanoscale Materials Behavior—Why all the Fuss? Presented by: Mark A. Ratner, Ph.D., Department of Chemistry, Northwestern University. http://www.blueskybroadcast.com/Client/ARVO/

National Cancer Institute: Video Journey Into Nanotechnology. http://nano.cancer.gov/resource_center/video_journey_qt-low.asp

Bones that Grow Back Video. http://www.sciencentral.com/

Careers in Nanotechnology Information Video. Penn State U. http://www.cneu.psu.edu/nePublications.html

Electrostatic Self-Assembly. NanoSonic. http://www.nanosonic.com/schoolkits/schoolkitsFS.html

NanoManipulator: Seeing and Touching Molecules. http://www.nanotech-now.com/multimedia.htm

Smaller than Small, National Geographic Magazine. http://www7.nationalgeographic.com/ngm/0606/feature4/multimedia.html

Nanozone, Lawrence Hall of Science. http://www.nanozone.org/whatvideo.htm

Clean Technology Vehicles. Altairnano. Demonstration vehicle using NanoSafe long-term batteries for vehicles. http://www.altairnano.com/ZEV.mov

Vega Workshop Videos. There are several videos in this collection. They include Harry Kroto's Buckyball Workshops and John Murrell's States of Matter Workshop. http://www.vega.org.uk/schools/details/4

Molecular Motors Drive Cellular Movements. A group led by Susan P. Gilbert, associate professor of biological sciences, is figuring out how motors inside cells convert chemical energy into mechanical force. http://www.pitt.edu/research.html

Nano-optics Research Offers a Bright Future. Researchers led by Hong Koo Kim, codirector of the Institute of NanoScience and Engineering, have developed a new technology that may revolutionize optics and fields like imaging, spectroscopy, and information technology. http://www.pitt.edu/research.html

Project on Emerging Nanotechnologies. "Hands-on" Learning Activity for Science Invisible to the Naked Eye. Bethany Maynard, a 6th grader at a Fairfax County,

Virginia elementary school, shows how young people can observe, test, and investigate nanotechnology. http://www.nanotechproject.org/76/nanotechnology-can-be-childs-play

Nanowires and Nanocrystals for Nanotechnology. Yi Cui is an assistant professor in the Materials Science and Engineering Department at Stanford. video.google.com/videoplay?docid=6571968052542741458

Synthesis of Colloidal Silver. The formation of silver nanoparticles can be observed by a change in color, small nanoparticles of silver are yellow. http://mrsec.wisc.edu/Edetc/nanolab/silver/index.html

The Lemelson Center. What is a buckminsterfullerene? Sir Harold Kroto. The Nobel Laureate explains why he named the carbon cluster that he discovered as a buckminsterfullerene. http://www.invention.smithsonian.org/video/ and http://invention.smithsonian.org/centerpieces/ilives/kroto/kroto.html

Dr. Alan Goldstein on potential and dangers of nanotechnology. http://video.google.com/videoplay?docid=5221560918013409256&hl=en

Cellular Visions: The Inner Life of a Cell, the animation illustrates unseen molecular mechanisms and the ones they trigger, specifically how white blood cells sense and respond to their surroundings and external stimuli. The animation shows a number of molecular machines—ribosomes, motors, and more. http://www.studiodaily.com/main/searchlist/6850.html

Probe Microscopes. Wendy Crone. Activities using magnetic probe strips to investigate the north and south poles of a magnet. Scanning Probe microscopy works the same way. Conversations in Science. Madison Metropolitan School District. UW-Madison Interdisciplinary Education Group. http://mrsec.wisc.edu/Edetc/cineplex/MMSD/scanning1.html

Nanotech Assembly. Productive Nanosystems: From Molecules to super products. Mark Sims, and Nanorex, Inc. http://singularityvideos.blogspot.com/2006/09/nanotech-assembler.html

Demonstration Video. See How It Works. Self-Cleaning Glass from Pilkington. http://www.pilkingtonselfcleaningglass.co.uk/howitworks;jsessionid=450BA85300D73FF706160BEECB8A1614

Nanotechnology Size and Scale. Professor Wendy Crone. Conversations in Science. Madison Metropolitan School District. UW-Madison Interdisciplinary Education Group. http://mrsec.wisc.edu/Edetc/cineplex/MMSD/nano3.html

N is for Nanotechnology. NISE Network. "N" is for Nanotechnology" is a 30-minute documentary exploring the hypes, hopes, and facts of this fascinating field as seen through the eyes of award-winning scientists, industry leaders, and writers. http://www.knhproductions.ca/nisnano/trailer.html

mPHASE Nanobattery. MPhase develops new nanobattery technology. Applications include homeland security, space exploration, communications, and the medical field. YouTube. http://www.youtube.com/watch?v=P8UwBP4yVgM

Nanotechnology Applications: Professor Lisensky shows some nanotechnology examples such as ferrofluids and carbon nanotubes and how these can be used to make displays, space elevators, cancer treatments, self-cleaning windows, and stain-free. Nano Quest Video Library. http://fll.ee.nd.edu/index.cgi?videoname=apps

Space Elevator. Scientists envision a space elevator based in the Pacific Ocean and rising to a satellite in geosynchronous orbit. http://www.sciencentral.com/articles/view.php3?article_id=218392162&language=english

NNIN Soft Lithography Network. Professor George M. Whitesides. This is a technical forum on how to use lithography in a small business as users and providers. http://www.cns.fas.harvard.edu/research/cns_videos.php

Photovoltaics: Turning Sunlight Into Electricity. United States Department of Energy. Solar Energies Technologies Program: Animations. http://www1.eere.energy.gov/solar/video/pv3.mov

Measuring Electrical Properties with an Electron Force Microscope. Professor Wendy Crone. Madison Metropolitan School District. http://mrsec.wisc.edu/Edetc/cineplex/MMSD/scanning3.html

What are the Symptoms of Diabetes? Who is at Risk? How is Diabetes Confirmed? Diabetes Clinic. IrishHealth. (56K/Dialup. http://www.irishhealth.com/clin/diabetes/video.html#

How a Fuel Cell Works. Ballard Fuel Cells. Ballard's principal business is the design, development, and manufacture of proton exchange membrane (PEM) fuel cell products. http://www.ballard.com/be_informed/fuel_cell_technology/how_the_technology_works

How Fuel Cells Work. A HowStuffWorks Video. Professor Tom Fuller from the Georgia Tech Institute of Technology explains how a fuel cell works. http://videos.howstuffworks.com/fuel-cell-video.htm

Hydrogen Fuel Cell. Digital Splash Multimedia Studios. http://www.digitalsplashstudios.com/fuel-cell.html

Detecting Toxics. A new portable lab that detects deadly chemicals in the air. ScienceDaily. http://www.sciencedaily.com/videos/2006-02-09/

Nature: The First Nanotechnologist. Dr. Gentry explains how nature works at the nanometer size scale to create the cells in our bodies; examples include cell walls, molecular self-assembly, and odor receptors. http://fll.ee.nd.edu/index.cgi?videoname=nature

Probe Microscopes: Tools used in Nanotechnology by Professor Wendy Crone. What are the tools that are used to explore the world of nano science? Professor Crone uses a common refrigerator magnet to illustrate a probe device for identifying atoms. The presentation continues with an illustration how nanoprobes investigate surfaces and move atoms. http://fll.ee.nd.edu/index.cgi?videoname=tools

Development of a platform based on a fleet of scientific instruments configured as autonomous miniature robots capable of fast operations at the nanometer scale. NanoRobotics Laboratory. Laboratoire de nanorobotique. Go to: http://wiki.polymtl.ca/nano/index.php/Research Then click on the site. New Micro-Nanorobotics Platforms.

Golf Gear Improving Your Score With Technology. AOL Video. http://video.aol.com/video-search/Golf-GearImproving-your-score-with-technology/id/3305225634

E. Coli Hand-Held Sensor. Detecting Bacteria With Electromechanical Cantilevers. Chemical engineers have developed a sensor that can almost instantly detect the presence of *E. coli.* Science Daily. http://www.sciencedaily.com/videos/2006-11-09/

Safer Water Worldwide. Industrial Toxicologists Develop Cost-Effective, Life-Saving Disinfection. Toxicologists developed PUR, a water purifier that combines a flocculant, which separates particles and organisms from water Science Daily. http://www.sciencedaily.com/videos/2006-12-06/

The Threat of Bird Flu. ScienceDaily. http://www.sciencedaily.com/releases/2007/04/070416092206.htm

Protein Lab Chip Assay. Agilent. A lab-on-a-chip used for analyzing proteins. http://www.chem.agilent.com/scripts/generic.asp?lPage=1566&indcol=N&prodcol=Y

Smart Trash Cans. A microchip embedded in recycling bins encourages households to recycle trash for dollars. Science Daily. http://www.sciencedaily.com/videos/2006-10-01/

Protein Lab Chip Assay. A lab-on-a-chip used for analyzing proteins. http://www.chem.agilent.com/scripts/generic.asp?lPage=1566&indcol=N&prodcol=Y

AUDIOS

Voyage of the Nano-Surgeons. NASA-funded scientists are crafting microscopic vessels that can venture into the human body and repair problems—one cell at a time. http://science.nasa.gov/headlines/y2002/15jan_nano.htm

The Lure of Nanotechnology. National Public Radio. http://www.npr.org/templates/story/story.php?storyId=1604431

"Buckyball" Nobel Laureate Richard Smalley Dies. http://www.npr.org/

Next Generation of Drug Delivery. The Bourne Report. There are all sorts of ways to get medicine into the body; here are a few examples of how MEMS and Nanotech-based approaches are making a difference. Marlene Bourne.

http://bournereport.podomatic.com/entry/2006-12-10T13_46_12-08_00

Appendix E

Some Important Events in Nanotechnology History

1900	Max Planck proposes energy quantization.
1905–1930	Development of quantum mechanics.
1927	Heisenberg formulated his uncertainty principle
1931–1933	The first electron microscope was built by Ernst Ruska and Max Knoll.
1953	DNA structure discovery by James D. Watson and Francis Crick.
1959	Feynman's talk, "There is plenty of room at the bottom."
1968	Alfred Y. Cho and John Arthur of Bell Laboratories and their colleagues invent a technique that can deposit single atomic layers on a surface.
1965	Proposal of Moore's Law.
1974	Norio Taniguchi conceives the word "nanotechnology" to signify manufacturing products with a tolerance of less than a micron.
1981	Invention of STM by Gerd Binnig and Heinrich Rohrer. The scanning tunneling microscope can image individual atoms.
1983	Ralph Nuzzo and David Allara of Bell Laboratories discover self-assembled monolayers. Nuzzo and Allara's research lead to the development of stain-repellent coatings on carpet, lubricants that still cling in harsh weather, and materials that line artificial hearts and keep the body's proteins from depositing.
1985	Buckyball discovery by Kroto, Curl, and Smalley.
1986	Invention of Atomic Force Microscope by Binnig, Quate, and Gerber
1986	Eric Drexler writes *Engines of Creation* that describes the manufacture of Nanoscale devices.
1989	Donald Eigler of IBM writes the letters of his company using individual xenon atoms.

1991	Discovery of carbon nanotubes by Sumio Iijima.
1996	Robert Curl Jr., Richard Smalley, and Harold Kroto receive a Nobel Prize for the discovery of buckminsterfullerene, the scientific name for Buckyballs.
2000	The Clinton administration announces the National Nanotechnology Initiative (NNI), which funds 700 million dollars a year.
2002	U.S. Army awards a contract to Massachusetts Institute of Technology to develop military applications for nanotechnology.
2003	President Bush Signs Nanotechnology Research and Development Act, which authorizes funding for nanotechnology research and development (R&D) over 4 years, starting in FY 2005.
2004	Carbon nanotubes used a light filament. President Bush Signs Bill Authorizing $3.7 Billion Nanotechnology Program for nanotechnology R&D, for FY 2005–2008.
2005	Beam of electrons used to shape metallic nanowires.
2006	Technology for making thin-film nanotubes by evaporation is invented. Food and Drug Administration faces a growing number of nanotech medical devices to evaluate.
2007	Russia decides to invest 1 billion dollars in nanotechnology from their funds in oil and gas reserves. Regulatory oversight of nanotechnology is urgently needed and the Environmental Protection Agency (EPA) should act now, reports an EPA study in 2007.

Appendix F

National Science Education Standards, Content Standards

Unifying Concepts and Processes, K–12
Systems, order, and organization
Evidence, models, and explanation
Constancy, change, and measurement
Evolution and equilibrium
Form and function

Science as Inquiry, Content Standard A, Grades 9–12
Abilities necessary to do scientific inquiry
Understandings about scientific inquiry

Physical Science, Content Standard B, Grades 9–12
Structure of atoms
Structure and properties of matter
Chemical reactions
Motions and forces
Conservation of energy and increase in disorder
Interactions of energy and matter

Life Science, Content Standard C, Grades 9–12
The cell
Molecular basis of heredity
Biological evolution
Interdependence of organisms
Matter, energy, and organization in living systems
Behavior of organisms

Earth and Space Science, Content Standard D, Grades 9–12
Energy in the earth system
Geochemical cycles
Origin and evolution of the earth system
Origin and evolution of the universe

Science and Technology, Content Standard E
Abilities of technological design
Understandings about science and technology

Science in Personal and Social Perspectives, Content Standard F
Personal and community health
Population growth
Natural resources
Environmental quality
Natural and human-induced hazards
Science and technology in local, national, and global challenges

History and Nature of Science, Content Standard G
Science as a human endeavor
Nature of scientific knowledge
Historical perspectives

Appendix G

Colleges and Museums

National Nanotechnology Infrastructure Network (NNIN) has provided the following links as resources for nanotechnology education. These links are arranged by institution to which these educational and outreach programs are affiliated. Some links are devoted to middle school and high school students, while a majority contains information for all levels. A brief description for each link is provided.

Albany NanoTech: Albany NanoTech at the University at Albany—SUNY brings together the nanoelectronics, nanosystems, and nanophotonics technologies that power the nanotechnology revolution. http://www.albanynanotech.org

Arizona State University: Includes images of animations, microscopy, schematics, and lecture videos in engineering, life sciences, math, technology, and physical sciences. Also includes interactive activities on carbon allotropes, biominerals, engineered materials, liquid crystals, yeast, iridescence, modern information storage media, gold films, music of spheres, light bulb, friction, DNA, etc. http://invsee.asu.edu/Invsee/invsee.htm

Columbia University: Details on seminars, NanoEngineering Clinic at Rowan, nanotechnology courses and workshops, REU, RET, and NanoDay in New York. Details on the REU, RET, High School Visitation & Ron McNair CITIES programs.
http://research.radlab.columbia.edu/nsec/outreach/
http://www.cise.columbia.edu/mrsec/education.htm

Cornell University: Lesson plans on cantilevers, diatoms, microscopy and scale, motors, sunprint paper and photolithography, sodium alginate polymers, chromatography, dissolving chocolate, springboks, create a mechanical flea, size, catapults, elasticity testing, frog jumping, paper thickness, immunology curriculum, nanosmores and photolithography, and microrebus. http://www.nbtc.cornell.edu/mainstreetscience/index2.html

Lehigh: ImagiNations. Lehigh University Web site for students and teachers who want to know more about nanotechnology. http://www.lehigh.edu/~inimagin/

Museum of Science at Boston: Articles, news clips, demos, presentations, experiments, videos, speeches, etc., on nanotechnology. http://www.mos.org/doc/1137?words=nanotechnology

NanoScience Instruments: Details on RET & REU programs at Virginia Tech, Cornell University, Harvard University, University of Wisconsin-Madison, Rice University, Northwestern University, University of Connecticut, University of California at Santa Barbara, Columbia University, and University of South Carolina. http://www.nanoscience.com/education/nanoschools.html

National Nanotechnology Infrastructure Network: This site provides general information about nanotechnology, links to additional resources, information on REU and RET programs, and an online science magazine for students. http://www.nnin.org.

North Carolina State University: In this unique software you explore characteristics of viruses with the use of atomic force microscopy to figure out what is making a puppy ill. Watch a phage invade a bacterium. Learn about how viruses take over a cell and multiply. Examine different virus structures and learn about the history of virology research. http://ced.ncsu.edu/nanoscale/materials.htm

Northwestern University: Materials World Modules and national Center for Learning and Teaching in Nanoscale Science & Engineering includes instructional materials, workshops, professional development, and video broadcasts
http://www.materialsworldmodules.org/
http://www.nsec.northwestern.edu/education.htm http://www.nclt.us

Rice University: Series of 12 self-contained nanoscale science and technology lessons. Twenty-minute Proof-of-Concept DVD—3D animated video combining two lesson/adventures: Welcome to the NanoLoft & DNA the Blueprint of Life? Interactive digital student workbook features the Research Laboratory, the NanoLoft, the DNA room, & the Nanotechnology room with information, exercises, games, sound bites, out-of-the-box imagination, songs, etc. nanokids.rice.edu/

ScienceCentral, Inc: Lesson plans on nano cancer fix, nano's downside, smallest robot, etc. http://www.sciencentral.com/

University of California at Berkeley: Includes researcher presentations, interactive demos and facilitated activities, physical/virtual contextual bridge exhibits, insights into research, human map of research and talking techno heads.
http://scienceview.berkeley.edu/wor/Exhibits.html
http://scienceview.berkeley.edu/wor/Partners.htm

University of Chicago: Includes details on research opportunities for undergraduates and high school students, demos for elementary school kids and internships for high school students. Also, includes videos like "Sights & Sounds of Science," "DrangonflyTV," etc. Talks about museum collaborations & science camps. http://mrsec.uchicago.edu/outreach/

University of Illinois at Urbana-Champaign: The Bugscope project provides a resource to classrooms so that they may remotely operate a scanning electron microscope to image "bugs" at high magnification. The classroom has ownership of the project—they design their own experiment and provide their own bugs to be imaged in the microscope. http://bugscope.beckman.uiuc.edu/

University of North Carolina at Chapel Hill: Good site to explore relative sizes of objects. Contains the classic video "Powers of 10" by Eames & Eames, along with other resources for investigating scientific notation and the scale of things. Interactive Web site that starts with a pinhead & scales down to a virus. http://www.cs.unc.edu/Research/nano/ed/scale.html

University of South Carolina: The National Science Foundation awarded the Department of Chemistry & Biochemistry with Research Experiences for Undergraduates Program & Research Experiences for Teachers Program in Nanoscience. http://nano.sc.edu/ret/about/home.html

University of Wisconsin-Madison: Modules designed to show how x-ray diffraction & scanning probe microscopy, shape-memory alloys, light-emitting diodes, ferrofluids, magnetism, curricular connections, memory metal, and other metal nanoparticles illustrate basic science concepts covered in the traditional chemistry curriculum. http://www.mrsec.wisc.edu/edetc/modules/index.html

Vega Science Trust, England: Videos on basics of nanotechnology and how it will change the world. http://www.vega.org.uk/series/tnbt/nanotechnology/index.php

Virginia Tech: First edition of Nano2Earth, a secondary school curriculum designed to introduce nanoscience and nanotechnology. It is the first program in the country to introduce these subjects using an environmental science approach. http://www.nanoed.vt.edu/curriculum2.htm

Glossary

adhesion. The force causing unlike molecules to be attached to one another.

alkali metals. A group of soft, very reactive elements that include lithium, sodium, and potassium.

angstrom unit. One hundred-millionth of a centimeter or 10^8.

anion. A negatively charged ion.

arsenic. A gray, brittle element with a metallic luster.

assay. A chemical test to determine the effect of a drug.

atom. The smallest unit of a chemical element, about a third of a nanometer in diameter. A unit composed of neutrons, electrons, and protons. Atoms make up molecules and solid objects.

Atomic Force Microscope (AFM). An instrument able to image surfaces by measuring the force on a tip as it moves across a surface on a substrate. Also termed a scanning force microscope.

biopolymer. A polymer found in nature. DNA and RNA are examples of naturally occurring biopolymers. See also polymer.

biosensor. A sensor used to detect a biological substance (for example: bacteria, blood gases, or hormones).

biotech. Biotechnology, a technology based on biological organisms or molecular biological techniques.

bottom-up nanofabrication. Building larger objects from smaller building blocks such as atoms and molecules.

Buckminsterfullerene. *See Fullerenes.* A broad term covering the variety of buckyballs and carbon nanotubes that exist. Named after the architect Buckminster Fuller, who is famous for the geodesic dome, which buckyballs resemble.

Buckyball. A large molecule made up of 60 carbon atoms arranged in a series of interlocking hexagonal shapes, forming a structure similar to a soccer ball.

carbon. A nonmetallic element found in all living things. Carbon is part of all organic compounds and, in combined form, of many inorganic substances. Diamonds, graphite, and fullerenes are pure forms of carbon.

carbon nanotubes. Long, thin cylinders of carbon, that are unique for their size, shape, and remarkable physical properties. Nanotubes have a very broad range of electronic, thermal, and structural properties that change depending on the different kinds of nanotube length, chirality, or twist.

catalyst. Any substance that increases a chemical reaction without itself being consumed by the reaction.

cell. A small, usually microscopic, membrane-bound structure that is the fundamental unit of all living things.

cholesterol. A large molecule found in living tissue including much of the mass of the human liver.

cleanroom. A climate- and particle-controlled workspace that includes an air filtration system that changes the air in the cleanroom about ten times every minute. Special cleanroom suits are also required to protect equipment and other materials in the cleanroom.

compound. A material in which atoms of different elements are bonded together.

crystals. The formation of a solid whose atoms have a definite pattern or arrangement.

dendrimer. A dendrimer is a polymer with physical characteristics that make it very applicable to probe and diagnose diseases or to manipulate cells at the nanoscale. Dendrimer comes from the Greek word *dendra*, meaning tree.

Dip Pen Nanolithography. An AFM-based soft-lithography technique. A method for nanoscale patterning of surfaces by the transfer of a material from the tip of an atomic force microscope onto the surface.

DNA (deoxyribonucleic acid). The molecule that encodes genetic information, found in the cell's nucleus.

DNA Chip. A purpose-built microchip used to identify mutations or alterations in a gene's DNA.

drug delivery. The use of physical, chemical, and biological components to deliver controlled amounts of a therapeutic agent to a diseased cell.

E-beam. An electron beam focused, steered, and controlled by magnets and by electrostatic lenses, such as in an e-beam writer or a Scanning Electron Microscope (SEM).

electrode. A material that allows an electric current to enter or leave a device.

electron. A subatomic particle with one negative charge.

electron beam lithography. A process of fabrication that uses electron beams to form structures on surfaces.

electron microscopy. An electron microscope uses electrons rather than light to create an image. An electron microscope focuses a beam of electrons at an object and detects the actions of electrons as they scatter off the surface to form an image.

element. A material consisting of only one type of atom.

fuel cell. An electrical cell that converts chemical energy of a fuel into direct-current electrical energy. Researchers are hoping to develop fuel cells that could take the place of combustion engines, thereby reducing the world's dependence on fossil fuels.

ferrofluid. A fluid in which fine particles of iron, magnetite, or cobalt are suspended, typically in oil. Ferrofluids were invented by NASA as a way to control the flow of liquid fuels in space.

ferromagnetic materials. Substances, including a number of crystalline materials that are characterized by a possible permanent magnetization.

fluorescence. The property of molecules to absorb a wavelength of light and then emit light at a higher wavelength.

fullerenes. Fullerenes are a molecular form of pure carbon discovered in 1985. The most abundant form produced is buckminsterfullerene (C60), with 60 carbon atoms arranged in a spherical structure. There are larger fullerenes containing from 70 to 500 carbon atoms.

in vivo. A medical experiment done within a living subject.

ion. An atom or molecule that is electrically charged.

ion conductors. The discharge of charged particles in a fluid electrolyte to conduct an electrical current.

ionic bond. A chemical bond in which an attractive electric force holds ions of opposite charge together.

lab-on-a-chip devices. Miniaturized analytical systems that integrate a chemical laboratory on a chip. Lab-on-a-chip technology enables portable devices for point-of-care (or on-site) medical diagnostics and environmental monitoring.

liposome. A type of nanoparticle made from lipids or fat molecules. It was the first nanoparticle used to create therapeutic agents to treat infectious diseases and cancer.

lithography. The process of imprinting patterns on materials.

logic gates. Fundamental logic structures that are used in digital computing. The most common gates are AND, OR, NOT Logic gates.

Magnetic Force Microscope. A kind of scanning probe microscope in which a magnetic force causes the tip to move. The motion allows the operator to measure the magnetic force of a sample.

matter. Anything that occupies space.

metastasis. The process by which certain cancers spread from one organ or structure within the body to another.

microchip. A silicon chip that contains many microscopic components.

molecular manufacturing. The automated building of products from the bottom up, molecule by molecule, with atomic precision. This will make products that are extremely lightweight, flexible, and durable.

molecular motors. Nanostructures that work by transforming chemical energy to mechanical energy within biological structures.

molecules. Molecules are groups of atoms, which are bonded together.

monomer. A small molecule that may become chemically bonded to other monomers to form a polymer. From the Greek mono "one" and meros "part."

Moore's Law. Coined in 1965 by Gordon Moore, future chairman and chief executive of Intel, it stated at the time that the number of transistors packed into an integrated circuit had doubled every year since the technology's inception four years earlier. In 1975 he revised this to every 2 years, and most people quote 18 months.

nanobiotechnology. The ability to develop the tools and processes to build devices for studying biosystems, in order to learn from biology how to create better nanoscale devices.

nanocomposites. Nanomaterials that result from the mixture of two or more nanoparticles to create greater strength in a product.

nanocrystals. Nanocrystals are aggregates of thousands of atoms that combine into a crystalline form of matter. Typically around 10 nanometers in diameter, nanocrystals are larger than molecules but smaller than bulk solids. The crystals might be added to plastics and other metals to make new types of composite structures for everything from cars to electronics.

nanodots. Nanoparticles that consist of homogenous materials that are spherical or cubical in shape.

nanofabrication. The construction of items using assemblers and stock molecules.

nanofiber. A polymer membrane formed by electrospinning, with filament diameters of 150–200 nanometers.

nanolithography. Refers to etching, writing, or printing at the microscopic level, where the dimensions of characters are on the order of nanometers.

nanomanipulation. The process of manipulating items at an atomic or molecular scale in order to produce precise structures.

nanomedicine. The area of research focusing on the development of a wide spectrum of nanoscale technologies for disease diagnosis, treatment, and prevention.

nanometer. A unit of measurement equal to one-billionth of one meter. The head of a pin is about 1 million nanometers across.

nanoparticle. A nanoscale spherical or capsule-shaped structure. Most nanoparticles are hollow, which provides a central reservoir that can be filled with anticancer drugs, detection agents, or chemicals. Most nanoparticles are constructed to be small enough to pass through blood capillaries and enter cells

nanoprobe. Nanoscale machines used to diagnose, image, report on, and treat disease within the body.

nanorods or carbon. Formed from multiwall carbon nanotubes.

nanoscale. The length scale between 1 to 100 nanometers.

nanoshells. Nanoscale metal spheres, which can absorb or scatter light at virtually any wavelength. Nanoshells are being investigated for use in treating cancer.

nanoscience. The scientific understanding of nanoscale.

nanospheres. Spherical objects from tens to hundreds of nanometers consisting of synthetic or natural particles.

nanostructures. Structures whose overall design is at the nanoscale.

nanotechnology. A manufacturing technology to fabricate most structures and machines from individual atoms and molecules.

nano-test-tubes. Carbon nanotubes that are opened and filled with materials, and used to carry out chemical reactions.

nanotube. A one-dimensional fullerene with a cylindrical shape. Carbon nanotubes were discovered in 1991 by Sumio Iijima. Nanotubes are a proving to be useful as molecular components for nanotechnology.

nanowires. Semiconductor nanowires are one-dimensional structures, with unique electrical and optical properties, that are used as building blocks in nanoscale devices.

neutron. A subatomic particle with no electrical change and positive charge.

optics. The science of light and its interaction with matter.

piezoelectrics. Dielectric crystal that produces a voltage when subjected to mechanical stress or can change shape when subjected to a voltage.

photosynthesis. The process by which plants and bacteria transform energy from light sources into chemical energy.

photovoltaics. An artificial system that transforms light energy into electrical current.

polymer. A macromolecule formed from a long chain of molecules called monomers. Polymers may be organic, inorganic, synthetic, or natural in origin.

polymerization. The process of making polymers from monomers.

protein. Large organic molecules involved in all aspects of cell structure and function.

proton. A subatomic particle with a positive charge of one unit. The number of protons in a nucleus determines which element the atom is.

quantum. A small discrete package of light energy.

quantum dots. Nanometer-sized semiconductor particles, made of cadmium selenide (CdSe), cadmium sulfide (CdS), or cadmium telluride (CdTe) with an inert polymer coating. Researchers are investigating the use of quantum dots for medical applications, using the molecule-sized crystals as probes to track antibodies, viruses, proteins, or DNA within the human body.

quantum dot nanocrystals (QDNs). They are used to tag biological molecules.

quantum mechanics. A largely computational physical theory that describes the properties of matter at the nanometer scale.

replicator. A system able to build copies of itself when provided with raw materials and energy.

RNA (ribonucleic acid). A long linear polymer of nucleotides found mainly in the cytoplasm of a cell that transmits genetic information from DNA to the cytoplasm and controls certain chemical processes in the cell.

Scanning Force Microscope (SFM). An instrument able to image surfaces to molecular accuracy by mechanically probing their surface contours. Also termed an atomic force microscope.

Scanning Probe Microscopy (SPM). Experimental techniques used to image both organic and inorganic surfaces with atomic resolution. Includes atomic force microscopes and scanning tunneling microscopes.

Scanning Tunneling Microscope (STM). An instrument able to image conducting surfaces to atomic accuracy to reveal the structure of a sample.

self-assembly. At the molecular level, the spontaneous gathering of molecules into well-defined, stable structures that are held together by intermolecular forces.

semiconductor. A solid substance, such as silicon, whose ability to conduct electricity is less than metals. The semiconductor's ability to conduct electricity increases with rising temperatures.

silicon. A nonmetallic element that is widely used as a semiconductor for making integrated circuits.

solubility. The ability of a solute to dissolve in a given solvent.

spectroscopy. The science of using a tool to reveal the composition of a sample by measuring the light absorbed, scattered, and emitted by atoms or molecules resulting in a spectrum.

stent. An expandable wire mesh used in a medical operation to keep blood vessels open.

substrate. A wafer that is the basis for subsequent processing operations in the fabrication of semiconductor devices.

superconductor. An object or substance that conducts electricity with zero resistance.

thin film. A film one molecule thick; often referred to as a monolayer.

top-down fabrication. The process of making a nanostructure starting with the largest structure and taking parts away.

top-down nanofabrication. Produces nanometer-scaled devices from bulk materials by lithography techniques, which include photolithography, e-beam lithography, and nanoimprint, etc.

transistor. The basic element in an integrated circuit. It is an on/off switch that determines whether a bit is one or two.

Transmission Electron Microscopy (TEM). The use of electron high-energy beams to achieve magnification close to atomic observation.

wavelength. The wavelength of light is usually measured in angstrom units.

zeolite. A ceramic material built of aluminum oxide and silicon oxide with other materials.

INDEX

About the Author

JOHN MONGILLO is a noted science writer and educator. He is coauthor of *Encyclopedia of Environmental Science, Environmental Activists,* and *Teen Guides to Environmental Science,* all available from Greenwood.